DEACTIVATED WEAPONS

A COLLECTOR'S GUIDE TO
DEACTIVATED WEAPONS

TONY JEEVES

IAN ALLAN
Publishing

First published 1993

ISBN 0 7110 2056 6

Published by Ian Allan Ltd, Shepperton, Surrey; and
printed by Ian Allan Printing Ltd at their works at
Coombelands in Runnymede, England.

Cover: Courtesy of Worldwide Arms.

Contents

Acknowledgements

I am indebted to many people in the preparation of this book. Firstly to Les and Marita Rawlins of Worldwide Arms, not only for the generous help and assistance provided through their extensive knowledge and access to their stocks of deactivated weapons but also for the time and resources they expended ensuring that deactivation would become legally recognised.

To Herb Woodend and Paul Ellis of the MoD Pattern Room for allowing me to photograph items in their superb collection that I was unable to find elsewhere and to my friend Barbara Tischler for lending me the camera that allowed me to take acceptable photographs of difficult subjects.

Finally, but by no means least, to my family: my wife Jean for her unstinting support, endless cups of coffee and for keeping the collection dusted; my son Philip for sharing the interest; and my 'gundog' Charlie for patiently sharing the hours at the word processor.

My thanks also go to the many others, too numerous to mention, who have offered invaluable advice and assistance.

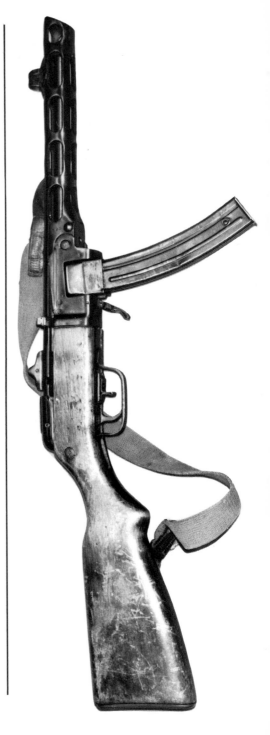

Foreword

There can be little doubt that the gun, in all its forms, has shaped the course of history and guided the destiny of nations more than any other device invented by man. It has helped to open up new territories, win independence from oppressors, change the outcome of wars, even build empires; it is, perhaps, therefore not surprising that guns hold a special fascination for hundreds of thousands of people around the world. However, history when it is viewed at a distance, through the pages of a book or even through the lens of a camera, may seem dull and uninteresting. The same history may be brought graphically to life when examples of weapons used during a crucial period are handled and examined.

For those fascinated by firearms, building a collection would seem to be the logical extension and manifestation of this fascination. However, there are many obstacles placed the path of those wishing to follow this course and gun collecting as a hobby is steadily being legislated out of existence. For older guns, such as matchlocks, flintlocks and percussion weapons, cost is usually the major barrier and where, as little as 20 years ago, good examples of many of these weapons could be picked up for very little, today even poor examples change hands for enormous sums. The laws of supply and demand rule in this market. The supply is limited as the weapons in question were built before mass production; and time has obviously taken its toll of those guns that were produced. Demand is great because in the majority of countries weapons of this age may be bought and displayed as antiques, without legal restriction.

For more modern weapons, the collector is often faced with a barrage of legislation, much of it Draconian in its power and the limitations it is able to impose on ownership. For instance, in the United Kingdom, any weapon using fixed ammunition, that is with cartridge and bullet in a single unit, requires a Firearms Certificate for purchase and possession. This in turn imposes the condition that the weapon be stored in a locked security cabinet, thus preventing satisfactory display and, as more than half the pleasure of collection of any item is in being able to display and handle that object freely, this is a major disadvantage.

Such legislation, ill-conceived and often hastily drafted, would seem to sound the death knell for the possibility of collecting any gun made since the late 19th century, as application for grant of the necessary permit, purely for the purpose of collecting such weapons would, at best, be greeted with obstruction by most police forces. To attempt to gain permission to collect automatic weapons such as machine and submachine guns, would reduce most local firearms officers to a state of helpless derision.

However, it is possible legally to own and collect such weapons and to display them in any way suitable. The answer is contained in the Firearms (Amendment) Act 1988. There is very little in the provisions of that Act that those with an interest in firearms could describe as positive. The vast majority of its provisions seem to be specifically designed to deprive gun enthusiasts of the few freedoms they had to enjoy their hobby. The one possible exception is Section 8, which deals specifically with the definition of legal deactivation of firearms. This Section effectively removes the previous 'Grey' areas of legislation and produces clear guidelines for deactivation of weapons to a universally acceptable standard.

It is true that deactivated weapons were available prior to this Act but, previously, no one was absolutely sure what was acceptable and there were 'horror' stories of collections of such weapons being confiscated. Now all weapons have to be deactivated in a specified manner and each weapon has to be accompanied by a certificate of deactivation, which may only be issued by an authorised Proof House after thorough examination of the weapon. The certificate provides absolute legal proof of the harmless status of the weapon.

This makes it possible for enthusiasts legally to buy, with no other licence or authority than the certificate of deactivation, a wide range of interesting weapons, ownership of which was previously either legally restricted or prohibited.

While exact methods of deactivation vary from weapon to weapon, deactivation *does not* mean welding the whole thing into a solid mass of useless scrap metal for, while it will be impossible for the weapon to be made to fire ammunition, the majority of the mechanism of the weapon will still function and it will remain an interesting and instructive collector's item. Deactivation also removes the weapon from the secure storage restrictions of the Firearms Act so that it may be displayed in almost any way the owner wishes.

For the collector, deactivation opens new possibilities. The weapons of the late 19th and 20th century — an interesting period that saw rapidly accelerating development — become readily available. Because weapons, particularly those of the 20th century, are not yet in particularly short supply, they are reasonably inexpensive, allowing whole collections to be built for a price that

would not purchase a single good example of a Colt percussion revolver. Effectively, we are at the beginning of a new stage in the evolution of the hobby of gun collecting.

The purpose of this book is to provide an introduction to the collection of modern deactivated military small arms. The potential for collection is enormous but I have chosen to limit this book to weapons of World War 2. This period of recent history is in itself fascinating and for the student of firearms, its rapid weapon development provides a rich variety of new types of small arms. Where the soldier of World War 1 fought brutal but relatively unsophisticated battles and had to learn little other than the use of a bolt action rifle, a single grenade and his bayonet; just over 20 years later, his son, taking part in another 'war to end wars', was expected to take greater interest in tactics and master a much greater array of weaponry.

The reason was that as the world approached the new conflict, many armies were still equipped with weapons that had seen service at the Somme or Gallipolli and were gearing training and strategies to the experiences of campaigns that were still very fresh in many memories. Gradually, as the nature of the forthcoming war became clearer, attitudes, tactics and weapons began to change and countries looked more closely at the weapons in their armouries.

As the war progressed, new weapons were designed and pressed into service sometimes because the performance of existing weapons was inadequate for the conditions of modern warfare; sometimes because the older weapons could not be produced in the quantities needed; or sometimes because updated tactics demanded updated weapons to provide an adequate response.

As a result, weapons were produced in greater variety and quantity than ever before and for this last reason, the majority are still in plentiful supply and therefore fairly inexpensive for the collector to obtain. This in turn makes it possible for the student of firearms, history or the collector of militaria to build an extensive collection of interesting weapons at moderate cost.

My intention is to cover those small arms that, at some time or other during World War 2, were the regular issued armament of recognised military formations of the main warring nations. It is however impossible to cover every weapon that was issued during this period; there were many weapons that were issued only to specialist formations; others from defeated nations that were incorporated into the victor's armouries and still more that were rescued from obsolete stores to fill gaps in production. Sometimes such weapons will find space in the following pages and I apologise in advance if the selection or omission may seem idiosyncratic to those readers with a special area of interest that they feel may have been neglected. I do hope you will however find the information contained here of interest and that it may encourage more people to discover the pleasure to be gained from collecting deactivated weapons.

1

The Law and Deactivation

The history of the hobby of collecting firearms is almost as long as the history of the weapons themselves, even back in the 14th century, the rich and powerful often amassed vast armouries of weapons captured from enemies and used them to decorate the walls of their castles and manor houses. For hundreds of years, the hobby remained limited to the privileged few; guns were expensive and the majority of people had more pressing demands on their available cash than gratification of a leisure interest. It was not until the 1920s, when Major H. Pollard published his book, the *History of Firearms*, that a more academic interest in the historic importance of firearms began to develop and, around this, as people had more leisure time and spare money, the collecting hobby evolved in a more general form. From that time the hobby escalated for the majority weapons available on the market were cheap, even by the price standards of the day and ownership was largely unhindered by licensing or other legal restrictions.

The gun collecting hobby then virtually travelled full circle to become once again the province of the privileged or lucky few. Price was the main deterrent for the majority of prospective collectors; for the types of guns, such as match, wheel, flint and percussion locks, that are currently obtainable without restriction, are very limited in number and therefore expensive. In an ideal world, people wishing to build a gun collection on modest means would be able to specialise in the more modern and therefore more plentiful and cheaper breech loading cartridge weapons produced this century or in the latter part of the

Below:
Verification that a weapon has been legally and irrevocably deactivated is provided by the Proof House Certificate that accompanies each gun. It details weapon type, serial number and country of origin.

Deactivation Certificate

issued by

The Birmingham Gun Barrel Proof House

Banbury Street, Birmingham B5 5RH

CAVENDO TUTUS

The Birmingham Proof House hereby certifies that work has been carried out on the firearm described below in a manner approved by the Secretary of State under Section 8 of the Firearms (Amendment) Act 1988 for rendering it incapable of discharging any shot, bullet or other missile.
No firearm certificate is required to possess this gun.

Type and Make: Sub-Machine Gun – Lanchester.

Number: 63808 Calibre/Chamber Length: 9mm.

Barrel Length: 8" Country of Origin: England.

Submitted by: Messrs World Wide Arms Ltd.

Certified by: Date: 13.11.1989

DA 6007

PROOF MASTER
for the Guardians of the Birmingham Proof House

Please note: (a) This certificate is an important document; it should be retained by the owner of the gun at all times.
(b) The main components of the gun to which this certificate relates have been marked with a Proof House inspection mark; these marks must not be removed or altered.

WARNING: Forging a de-activation certificate could constitute an offence under the Forgery and Counterfeiting Act 1981.

19th century. However, pursuing this course is far from easy for, over the years, more and more stringent laws have been imposed on the ownership of firearms to the stage where today the collection of most cartridge-firing small arms is, at the very best, 'frowned upon' by the authorities and in some areas police are under instruction from superiors actively to discourage private ownership of firearms. This attitude is greatly aided by the rather 'loose' wording of parts of the Firearms Act which allows it to be interpreted in a way that suits the prevailing attitude of the local Chief Constable and which leads to certain anomalies in administration and interpretation of the Act that almost invariably gravitate against the interests of the collector.

Live, as opposed to deactivated, small arms of World War 2 are covered by two Sections of the Firearms Act. Revolvers, semi automatic pistols and manually operated rifles are governed by Section 1 while machine guns, submachine guns, self-loading rifles and pistols which are capable of fully automatic or burst - fire come within Section 5.

Live Section 1 firearms may be legally owned by anyone in possession of an appropriately endorsed Firearms Certificate which may be issued to those with what is considered to be a legitimate reason for ownership. Such reasons usually included activities such as target shooting and membership of a recognised club. For the person only interested in collecting, the situation is even less clearly defined. It is possible to obtain a Firearms Certificate for the purposes of collecting but to be granted this the collector will need to convince the police that such a certificate should be granted and the onus will be very much upon the applicant to prove his bona fides as a collector to the authorities. This may not be easy, for such a person will not, like the target shooter, have the backing of a recognised club through which he will have proven his interest and abilities before making the application. If the collector persists, despite all the objections and obstacles that will be put in his way, and the application for a collector's Firearms Certificate is still refused, there is very little that can be done to appeal against the decision. Although it is theoretically possible to go through the courts, this can be a lengthy and costly process and success is far from guaranteed.

Even if a Certificate is granted for the purposes of collecting, it is very unlikely that permission for holding ammunition for the weapon will be given and it is certain that very strict limits will be placed on the number of weapons that may be possessed. The applicant will also be subject to the storage provisions of the Firearms Act and the police authorities will attempt to impose stringent security precautions, probably far in excess of those strictly required by the letter of the law. At the very minimum these will mean storing

weapons in a locked metal cabinet, securely fixed to the fabric of the building and may mean provision of an expensive alarm system and the storage of key parts of the weapons' mechanisms in separately secured conditions. The police will also have the power to conduct impromptu visits from time to time to check that the conditions of the certificate are being met and to inspect security. Penalties for any transgression are severe and the possession of even one round of live ammunition, contrary to the conditions of the licence, will invoke the full weight of the law and provide an excuse for confiscation of the collection and, quite probably, criminal prosecution.

The weapons falling within the Section 5 category are those described by the Firearms Act as 'specially dangerous' and as such they are prohibited from private ownership. Effectively these are all fully automatic firearms, from heavy machine guns through to the Mauser 'Schnellfeuer' pistol, as well as any rocket launcher or mortar projecting a stabilised projectile. Recent additions to this prohibited category are full-bore self-loading rifles, even if they were originally manufactured to be capable of only semi-automatic operation and shotguns under 41in in length. The sale or transfer of such weapons requires a special licence, issued only under the authority of the Secretary of State, to dealers and others able to meet extremely stringent special requirements.

The effect of the law is to put a large proportion of the military weapons of the 20th century, in live condition, totally out of reach of the military historian or collector of weapons and to make the remainder available only to those willing and able to accept the cost, inconvenience and loss of freedom involved in conforming to the required licensing and security conditions and prepared to forego the major pleasure to be gained from collecting any item — the ability to display and handle it with complete freedom.

Collecting live firearms can also be a risky business, for the collector is at the mercy of extremely volatile firearms legislation where what is currently legally collectable may suddenly become a prohibited weapon almost at the whim of the Home Secretary. A prime example is contained within the Firearms (Amendment) Act 1988 which decreed that all self-loading rifles over 0.22in calibre would fall within Section 5 of the Firearms Act and could thus no longer be legally owned. Many thousands of previously legally owned rifles became, almost overnight, prohibited weapons and as such, subject to confiscation. In return for weapons handed in, owners were given compensation which, in most cases, did not accurately reflect the value of the weapon. Who is to say that another change in the law may not similarly prohibit self-loading pistols or indeed any other type of weapon?

A collector of weapons prohibited in this way would have little or no alternative but to surrender his collection and accept whatever level of compensation (if any) that was offered. Such legal moves are far from impossible for a Parliamentary Green Paper has already been published that proposes a future prohibition on the collection of cartridge weapons and, while this proposal is as yet far from becoming law, the germ of the idea is obviously there and only requires the right pressure and conditions to cultivate it into fully-fledged legal force. Should this happen, appeal is unlikely to succeed as shooters, collectors and their organising bodies have already proven themselves woefully inept at opposing changes in the law. In the highly charged emotional atmosphere that currently surrounds firearms legislation, it is probably, therefore, only a matter of time before the collection of live cartridge-firing firearms becomes a thing of the past and huge numbers of interesting weapons are consigned to the melting pot. Once this particular ball begins to roll, there is no saying where it may end. Could even percussion arms, flint and wheel locks follow?

The collection of firearms changed dramatically with the 1988 Firearms (Amendment) Act which, for the first time, legally classified the collection of deactivated guns. Since then, deactivation has provided a possible answer for those keen to collect weapons, and enables them to do so virtually without restriction on the type of weapon or in the way in which the weapon is displayed and stored. It has also, once again, brought the hobby of collection full circle, putting large numbers of weapons within the reach of the average person of normal means.

Deactivation involves the alteration of key parts of the weapon in such a way that, as the Firearms Act states 'it has been rendered incapable of discharging any shot, bullet or other missile and has consequently ceased to be a firearm within the meaning of the Act'. There is a view, held by the 'purist' gun collectors, that deactivation is legalised vandalism and desecration and that weapons should be collected in original condition or not at all. In a Utopian society this could be considered to be a valid attitude but in the real world it is, at best, sanctimonious, for it ignores the fact that a large proportion of interesting modern weapons are, in any case, prohibited from private ownership and, as such, are destined for the scrap heap when their useful life is ended. Therefore, future generations will be unable to handle examples and only able to view them through the pages of books or in museums.

In effect, Section 8 of the Firearms (Amendment) Act 1988 — which specifically covers the legal deactivation of weapons — created a new category for gun collectors by specifying the actions necessary to transform any weapon into a 'non-firearm' and thus permanently remove it from the requirements of the firearms regula-

Above:
Key components are marked with special Proof House deactivation marks. The Birmingham Proof House uses crossed swords with the lettering 'DA' for deactivation and the last two numbers of the year.

Above:
The London Proof House has a scimitar with the alpha numeric code.

tions. As such, the weapon may be bought and owned without any form of licence and may be displayed, just like any other item of militaria, in whatever manner the owner wishes.

Prior to the drafting of the Amendment Act, deactivation had been possible but the guidelines for this had been very loose. It had been possible to buy military small arms either deactivated, without a licence, or, with the rifling ground out to smooth bore the barrel, to make purchase under what was then an easy to obtain Shotgun Certificate. There were no absolutely clear guidelines as to what the authorities considered to be adequate deactivation. Also, many military weapons, although smooth bored to become officially 'shotguns', were still capable of firing live ball ammunition. These factors led to legal problems and prosecution of some owners, as individual police forces attempted to enforce their own interpretation of the loose regulations.

The 1988 Firearms (Amendment) Act, largely drafted without consultation with representatives of shooters, was generally destructive to the shooting sport. An exception was in the field of deactivation. Here, Les Rawlins of Worldwide Arms decided to 'grasp the nettle' and commissioned Legal Counsel to analyse the strict letter of the firearms legislation and, armed with the facts that this provided, began extensive and well informed lobbying of the Home Office. This led, through a series of meetings, to full consultation in the drafting of the comprehensive set of guidelines for approved deactivation of all types of firearms that were eventually to form the basis of Section 8 of the Firearms (Amendment) Act.

The result was that the guidelines actually achieved the seemingly impossible task of satisfying the Government's need to be certain that weapons so treated would be irreversibly altered, while reconciling this with the collectors' natural

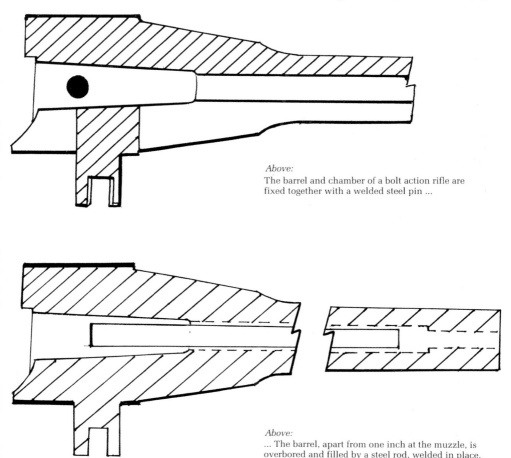

Above:
The barrel and chamber of a bolt action rifle are fixed together with a welded steel pin ...

Above:
... The barrel, apart from one inch at the muzzle, is overbored and filled by a steel rod, welded in place.

13

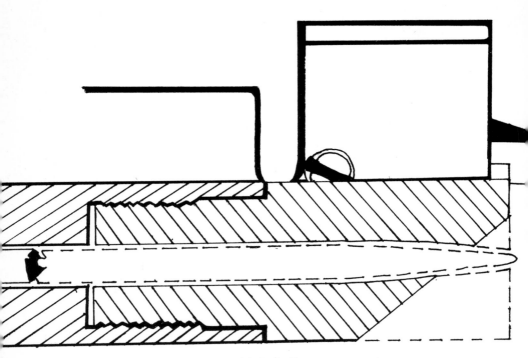

Above:
A section through the bolt showing chamfered bolt face and shortened firing pin.

desire for weapons that retained as many of their original functions as possible and that would not merely be hunks of inanimate scrap iron. It was also agreed that, wherever possible, methods of deactivation would be established that allowed the outward appearance of the weapon to remain the same as that of the live weapon. A further suggestion, made by Les Rawlins, was that each weapon should be inspected by either the Birmingham or London Proof House — normally responsible for certifying the safety of live weapons — and then issued with an individual Certificate of Deactivation. This certificate detailed the type, calibre and make of the weapon, its serial number, barrel length and country of origin. This was to serve as official recognition that the weapon had been deactivated in compliance with the Home Office guidelines and to provide incontrovertible evidence to police authorities anywhere that the weapon had indeed been made harmless in the required fashion. The certificate, signed by the Proof Master and numbered, would clearly state that the weapon had been correctly rendered inert and that a Firearms Certificate or other form of permit was no longer required for possession of the gun.

In addition to inspecting the weapon and providing the certificate, the Proof House also has the responsibility of stamping the key components of the weapon with special Proof Marks of deactivation. In the case of the London Proof

House this takes the form of a Scimitar with the letters 'DA' above and the last two digits of the year of deactivation below. For Birmingham, the mark consists of crossed swords with the letters and numbers stamped between the blades.

The Home Office guidelines allow the party actually carrying out the physical work of deactivation some latitude in the methods employed. In fact, for the collector, some methods of deactivation will be more acceptable than others and retain more of the original appearance. Also, a sensitive approach taken by the dealer in applying the guidelines will ensure that the actions necessary for deactivation will be conducted in a way that makes them as unobtrusive as possible while still complying with the law. For each of the main types of weapons, the Home Office guidelines are as follows:

Bolt Action Rifles
A slot, at least two thirds of the width of the bore diameter must be cut through the chamber and barrel, from the feed ramp to the full length of

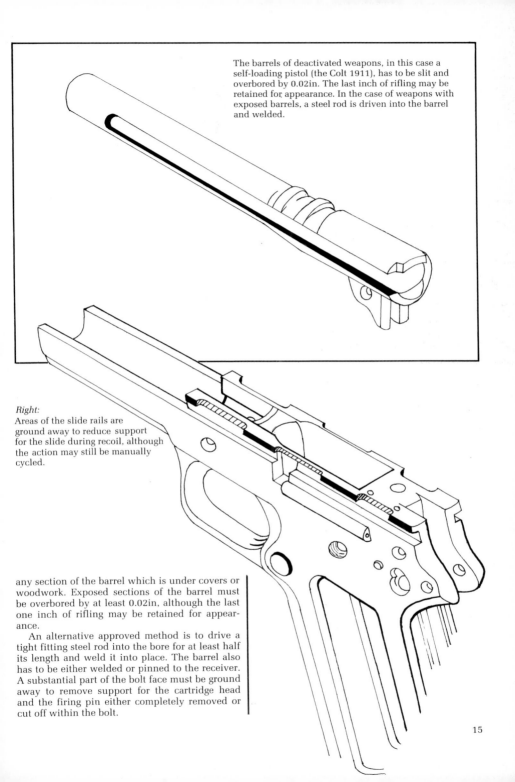

The barrels of deactivated weapons, in this case a self-loading pistol (the Colt 1911), has to be slit and overbored by 0.02in. The last inch of rifling may be retained for appearance. In the case of weapons with exposed barrels, a steel rod is driven into the barrel and welded.

Right:
Areas of the slide rails are ground away to reduce support for the slide during recoil, although the action may still be manually cycled.

any section of the barrel which is under covers or woodwork. Exposed sections of the barrel must be overbored by at least 0.02in, although the last one inch of rifling may be retained for appearance.

An alternative approved method is to drive a tight fitting steel rod into the bore for at least half its length and weld it into place. The barrel also has to be either welded or pinned to the receiver. A substantial part of the bolt face must be ground away to remove support for the cartridge head and the firing pin either completely removed or cut off within the bolt.

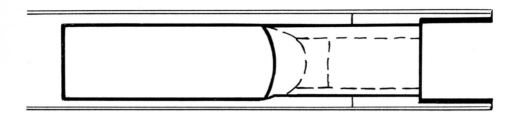

Self Loading Rifles, Machine Guns & Submachine Guns

Similar methods of treatment are employed for the barrels of these weapons as for bolt action rifles with either a slot at least two thirds of the width of the bore diameter cut through the chamber and barrel, from the feed ramp and running the full length of any section of the barrel which is under covers or woodwork. Again, exposed sections of the barrel must be overbored by at least 0.02in, although the last one inch of rifling may be retained for appearance. The alternative method of driving a tight fitting steel rod into the bore for at least half its length and welding it into place is also permissible and the barrel again has to be either welded or pinned to the receiver. For gas operated weapons, the gas piston must be removed.

Above:
A slot the width of the magazine well is cut through the feed ramp.

Below:
A revolver barrel has a tight fitting rod inserted and is held in the frame by a welded steel pin.

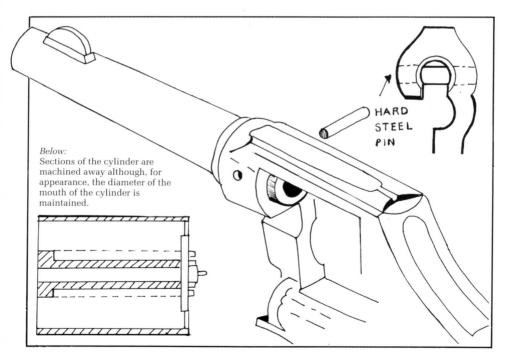

HARD
STEEL
PIN

Below:
Sections of the cylinder are machined away although, for appearance, the diameter of the mouth of the cylinder is maintained.

As far as the action is concerned, one method of treatment is replacement of the bolt, bolt carrier and springs from the receiver and their replacement with a thin metal tube of appropriate diameter welded into the receiver. This is not really satisfactory as it prevents any form of function for the action. The much better alternative is for the original bolt or bolt carrier to be retained but with a section, including part of the breech face, removed. In this way much of the action may still be cycled and field stripped.

Revolvers

The centre section of the cylinder has to be milled out in such a way that all chambers are opened up. However, a thin layer of metal can be retained around the axis pin and support retained at the front of the cylinder so that its external appearance remains largely unaltered.

A slot, at least one inch in length and two thirds the width of the bore, must be cut in the barrel which must be overbored by 0.02in although the last inch of rifling may be left for aesthetic purposes. Alternately, a tight fitting

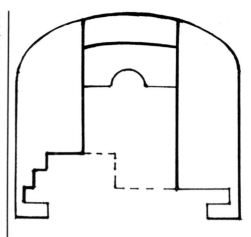

Above:
Dotted lines illustrate the sections of the breech block of a self-loading pistol which are removed.

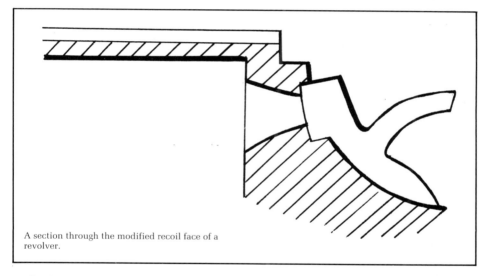

A section through the modified recoil face of a revolver.

steel rod, at least three quarters the length of the barrel may be driven into the bore and welded into place. The barrel must be permanently fixed to the frame with a hard steel pin welded in place. An area of the recoil face around the firing pin hole must be cut away to remove support for the cartridge head.

Self Loading and Machine Pistols

The striker must be removed or shortened and the recoil face and underside of the striker housing cut away. A proportion of the slide rails, suf-

ficient to reduce support for the slide during recoil in firing, must be removed and a slot the width of the magazine well cut through the feed ramp. If the ramp is an integral part of the barrel, a deep cut must be made into the frame close to the trigger.

The barrel must be treated in a way similar to that of the revolver with a slot, at least one inch in length and two thirds the width of the bore cut in the barrel which, apart from the last inch of rifling, must be overbored by 0.02in. Again, there is the alternative solution of driving a tight fit-

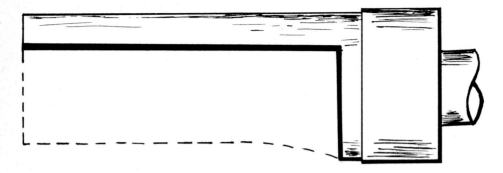

Above:
In blow back weapons, such as the Sten submachine gun, strategic parts of the bolt are removed.

ting steel rod, at least three quarters the length of the barrel, into the bore and welding it in place.

Mortars

For those mortars with a fixed firing pin, the firing pin and part of the breech must be machined away. For those with spring loaded firing pins, the pin together with any associated mechanism must be removed and part of the firing pin housing machined away. The breech block is welded to the main tube which has cross pins welded in place, a large steel ring welded close to the breech and weld material placed at points on the internal walls.

Rocket Launchers

All electrical and mechanical firing devices are removed and any covers for these mechanisms are welded into place. The launcher tubes are disabled in the same manner as those of mortars with welded cross pins and weld material and steel rings on the internal surfaces.

Most dealers will adopt the more 'sympathetic' approaches to the task of deactivation that are allowed by the guidelines but it is still worth querying the physical methods of deactivation employed when buying from a source for the first time. Wherever possible it is also worthwhile examining examples for, even where weapons are deactivated sympathetically, mechanical skill and a degree of craftsmanship are required to do this well.

2
On collecting

The motives for collecting guns are as varied as the weapons themselves. What it is that ignites that first spark of interest is a very personal thing. For me, the interest began as a child at the end of the war, playing 'Cowboys and Indians' with a friend in a wood behind his house, with a Mauser carbine and Walther pistol his father brought back as trophies from Dieppe. Then the world was full of guns; relatives and neighbours brought them home on leave and uncles in the Home Guard stored rifles in a corner of the kitchen. In that climate, no one expressed any horror that a pair of four or five year olds should be running around with real guns, albeit rendered relatively inanimate, by simple removal of parts. To a formative mind, the weight, feel and mechanism of the guns made them infinitely preferable to the crude and fragile toy guns of the time.

The interest was later reinforced by opportunities to use a variety of military weapons of many nations and was supplemented by a growing interest in military history and tactics which underlined the often crucial role well designed and employed small arms played in deciding the outcome of battles, campaigns and even wars. There was of course enormous frustration in being only able to handle many such weapons

under strictly controlled circumstances and never being able to own the majority personally or to study them at leisure or in real depth. The recent advent of legalised deactivation has provided an easy and relatively cheap way to overcome this frustration while not breaking the budget for, because the hobby is in its infancy, prices of even the more exotic 20th century weapons, are low in comparison to those current in other areas of gun collecting. There is every reason to believe that this branch of gun collecting will follow the pattern of others and that prices will escalate as the hobby grows and stocks of available weapons diminish. Therefore, starting to build a collection of deactivated weapons now could be a very good investment for the future. In the unlikely event that such an escalation of prices does not occur, the collector will still gain the pleasure of owning a small but dramatic piece of history with every gun added to the collection.

20

Asking where to begin collecting deactivated weapons is almost like asking how long is a piece of string. Many potential collectors will begin with an existing area of interest, whether it be in a particular type of weapon; weapons of a particular nation; weapons using a particular method of operation; those with which the collector has a personal association or those which will supplement an existing collection of militaria. Any of these factors may determine the starting point of a collection.

Whatever the motivation for collection, in the beginning it is worth postponing the purchase of the first weapon until you have a firm idea of the direction you wish the whole collection to take. While it is undoubtedly fun to go out and buy a gun, to treat the purchase in isolation can very easily be a later source of disappointment. Research is the key to satisfactory collection in any field. It is vital to read as much material as can be obtained about the types of weapon that interest you, from as many varied sources as possible and always try to see material that was originated as close to the origins of the weapon, both in terms of time and geography, as is practical. It is also wise not to take any one source as gospel for you will doubtless come across instances where writers have accidentally created an error or made an omission in a text. Sometimes this is later picked up and perpetuated by other authors until it becomes an almost accepted fact. By following the courses of action outlined above it will be possible to narrow and define your band of interest, to establish a structure to your collecting and build a display that will tell a story or which will follow a logical course of development and which will ultimately be much more interesting and satisfying. In itself the research will be an absorbing occupation and will possibly open up new and unexpected avenues of collection. It is not necessary to purchase an extensive (and expensive) library of source material. Local libraries normally house a good selection of the readily available books on guns and military history and these will form a sound foundation upon which to begin research. For more

Top right:
Some dealers in deactivated weapons keep enormous stocks. In this case Soviet Moisin Nagant rifles are pictured awaiting deactivation.

Right:
Deactivation requires the complete stripping of the weapon ...

advanced research, the reading rooms of institutions, such as the Imperial War Museum and the National Army Museum, will provide access to more esoteric source material such as rare published works as well as official documentation and records. Normally special arrangements will need to be made to register as an authorised reader but such registration is usually easy to obtain for bona fide research and provides access to material probably unavailable elsewhere. Local museums, both public and regimental, may also provide useful material.

The next and most necessary step at least for the majority of us who have limited means, is to sit down and work out just how much money you are prepared to spend on the hobby. For most people, price will be an important consideration and there is no doubt that collecting some

Left:
... and skill in machining in order to ensure that the work is done correctly but without destroying the appeal of the weapon.

Below:
Once deactivated, the weapons, together with Proof House certificates, are stored ready for despatch.

types of deactivated weapons will be considerably more expensive than others. There are several reasons for this, the first and foremost of which is supply and demand. While the supply of World War 2 weapons is plentiful, the collector of deactivated weapons in a field in which there is an established collectors' or shooters' market for live weapons, will be in competition with the live weapon collectors and will have to pay the inflated prices that have developed. For the types of weapons being discussed in this volume, this principle applies to the collection of rifles and pistols which may still be owned and shot by those granted the appropriate Firearms Certificate.

PISTOLS

Pistols, which are largely perceived as officers' weapons have accumulated their own brand of 'glamour' and their collection and use for recreation has long been popular. Glamour and popularity play an important part in establishing price levels. These elements are clearly reflected in the prices of such weapons on the deactivated market. Here, the price of a very ordinary but popular semi automatic service pistol such as a P'08 Luger will probably closely match that of a medium machine gun such as the Vickers. Another result of this long term popularity will

be that the best examples of the popular standard pistols and their variants will have already been bought by shooters and collectors and those that do become available will only be sold at a high premium price.

The 'glamour' element, while itself intangible, does have a very tangible effect on prices. This is why a relatively inefficient but attractive looking combat handgun, such as the Luger, can command prices that are multiples of the prices charged for the undoubtedly ugly but far more effective Webley service revolvers. Why one weapon should have such special appeal is difficult to define for it has many facets. In the case of the Luger it is partly because it is a German weapon and these have their own appeal to many collectors; partly because it looks pretty but mostly it is because an enormous mystique has developed around the Luger over the years that far outweighs its true significance in the annals of firearms development or military history. Although the Luger is an extreme example of the effect of this principle on prices, it is certainly not alone and, as a segment of the hobby of collecting deactivated weapons, the collection of pistols is the most expensive.

Nationality is another key factor in pricing. German, American and British weapons, of all types, are generally the most popular with collectors and therefore command higher prices than those of other nations, particularly weapons of countries in the Communist Bloc, which have yet to gain widespread popularity.

RIFLES

Rifles, although the majority (with the exception of automatic and semi automatic versions) are still available live, on licence to shooters and collectors, are affected to a much lesser extent by the glamour and hype elements. They are generally perceived as the 'workhorses' of armies and, as 'other ranks' issue, do not share the 'snob' appeal of pistols. Also, because they were issued in far greater numbers than pistols and there are fewer collectors of live weapons in this area, prices have not increased to anything like the same levels, remaining generally fairly cheap.

FULLY AUTOMATIC WEAPONS

Machine guns and submachine guns of World War 2 vintage are also relatively cheap on the deactivated market. The normal life of weapons of this era would be, firstly, issue to front line units of the warring nations and, once hostilities ceased, or the weapons became obsolete, they would go onto the surplus market to be sold on by the world's arms dealers to less developed nations before finally appearing for sale on the commercial market to civilians. In only rare exceptions is there a commercial civilian market for fully automatic weapons and so the majority become a drug on the market with little value. This is good for the collector of deactivated weapons for it tends to keep prices low although weapons in this category are still vulnerable to a degree of 'glamour inflation'. For instance, taking two contemporary and comparable weapons, the German MP40 and British Sten submachine guns as examples; the MP40 is perceived as a much more glamorous weapon, partly because it is a German weapon, partly because it has a more attractive design and partly because it seems to be better made than the unashamedly crude looking Sten. In fact both weapons were equally effective, produced in similar numbers and at only a slight difference in original cost to manufacture, yet on the deactivated market an MP40 will cost roughly four times the price of a MkII Sten.

HOW TO ACQUIRE WEAPONS

Although some dealers had been selling deactivated weapons for quite a few years before the advent of the 1988 Firearms (Amendment) Act, its clarification of guidelines for deactivation encouraged more traders to become involved in meeting the demand from potential collectors. As a result, the collector now has a wide choice of sources. The hobby of collecting deactivated weapons began with dealers selling via mail order and this method provides a convenient means of acquiring weapons and still accounts for the greatest volume of sales. The main benefit for the collector in buying from one of the dealer's catalogues is economy, for such dealers, with a high turnover of weapons, are able to buy in bulk and consequently acquire guns relatively cheaply and pass on the savings to customers. The best dealers spend a considerable amount of time scouring the world's markets for consignments of interesting items, occasionally unearthing 'gems' laying discarded and forgotten in warehouses in remote locations.

Although the mail order dealer specialises in items bought in bulk, it is still worth making contact and mentioning special interest in variations or special weapons as, sometimes, such items are included in consignments but do not figure in catalogues because they are not available in sufficient numbers to make this worthwhile. Also, if asked, the dealer will probably select weapons in extra fine condition if these are required by the collector and are available. In some cases it may also be possible to keep a special eye open for particular weapons in his travels and, if he knows he has a customer for an individual item, he may buy it where otherwise he would pass it by.

More high street gun retailers, bounded by laws and regulations on the majority of their stock, have welcomed the opportunity to trade in items they can hold and sell freely and have included deactivated weapons in their inventory. Often these weapons originate from the main mail order dealers but sometimes the retailer will purchase a small quantity of weapons and arrange for them to be deactivated himself and this can result in more unusual weapons, unavailable in sufficient quantities to interest the larger dealers, reaching the collectors' market. Whether or not the gunshop stocks deactivated weapons, it is worth looking at what live firearms are displayed and, if an interesting weapon is spotted, asking the dealer to have it deactivated. Remember, any weapon can be sent for deactivation and, while you will have to pay for the deactivation work and certification in addition to the cost of the weapon itself, this route does provide a means of adding unusual items to the collection or of filling a gap.

Similarly, but an area that has not yet, to my knowledge, been explored for deactivated weapons, is the purchase of guns at auction. Occasionally interesting modern military weapons, even fully automatic, appear in the auction catalogues. The options for the collector would be to either ask one of the dealers in deactivated small arms to bid on his behalf, or to place a bid directly through the auction house but first to make arrangements for the result of a successful bid to be passed to a dealer with the appropriate licensing for deactivation and certification. This would obviously require a fair amount of prior organisation and administration but the type of weapon we are concerned with in this volume normally sells quite cheaply at auction and the chances are that, even after allowing for various handling charges and the costs of deactivation, overall costs will remain low.

Many deactivated weapons are now to be found at the Arms Fairs which are held periodically at venues around the country. These attract not only the major dealers but also the smaller specialist outlets on whose tables may be found some of the more unusual weapons. The Fairs are held at all types of venues from the major exhibition complexes down to local recreation and social halls and they can be a treasure trove of riches for the collector of all types of militaria.

Some sectors of the shooting press act as good sources of information on the suppliers and stockists of deactivated weapons. Two publications are of particular interest and contain extensive advertising from suppliers and stockists of deactivated weapons as well as details of forthcoming auctions and arms fairs, these are: *Gunmart and Accessories* and *Guns Review*. Gunmart and Accessories also contains a good selection of classified advertisements from private sellers of deactivated weapons. Both magazines are published monthly and readily available from good newsagents.

Fakes and forgeries are not too much of a problem for the collector of World War 2 deactivated weapons. This is one of the advantages of starting at the beginning of a new hobby for we are currently at the stage with this hobby that the collecting of Third Reich militaria was, say, 20 years ago, when genuine material was so plentiful and cheap that it was not worthwhile for anyone to devote the time and effort necessary to manufacture a plausible fake. In the firearms field, counterfeiting of Colt revolvers and other high value antique firearms has become apparent. For 20th century weapons, such chicanery is extremely rare and is usually limited to the amendment of serial numbers and markings in niche areas of 20th century weapons collecting, such as that for Luger pistols, to pass off mundane models of weapons as more exotic and therefore expensive variants. No-one is manufacturing complete facsimile weapons of this era. What will be encountered will be copies of famous weapons such as the Thompson submachine gun, the Sten gun and the Mauser pistols, manufactured, usually in the Far East, either under licence (but more often unofficially) for domestic use. These were never made to deceive and are usually clearly identifiable either through their markings or by the poor standard of their manufacture. Such copies are collectable in their own right. One exceptional copy that does seem to have been made to deceive is the exceptionally rare 'Gerat Potsdam', 25,000 of which were manufactured, at tremendous expense, by Mauser at the tail end of the war. This weapon is a total copy of the MkII Sten, complete with forged British manufacturer's and inspector's marks, believed for issue for clandestine operations.

CONSERVATION

The majority of weapons from the era with which we are concerned and currently available on the deactivated market are in at least 'good used condition' and therefore seldom in need of restoration in its true sense. Most will show signs of use but to the collector these will be acceptable and may even enhance the attraction of the weapon. The main objective of the collector will be to ensure the weapon does not deteriorate while in his care even after years of storage and display. This can be ensured by taking a few simple steps.

The first stage will be to dismantle the weapon and procedures for this operation for most weapons can be found in books such as the invaluable Small *Arms of the World*. Where such

Above:
Ideas for display can be gained by studying how museums tackle the problem. This illustrates how the National Army Museum handles small displays.

information is not available, very often the procedure follows a logical progression which, once some basic knowledge of weapons of a particular type has been gained, can be worked out. If the weapon or procedure is unfamiliar, each step should be noted and the position of components roughly sketched so that they may be confidently returned. Stripping will reveal the true condition of the gun and allow cleaning. Old oil and hardened grease should be removed and this can be accomplished with a mild detergent solution and gentle brushing. Woodwork will often retain moisture and removing it will reveal patches of rust where it has been in contact with metal. The worst of heavy rust can be rubbed off with an oily rag and the remainder can generally be neutralised with proprietary rust treatment chemicals although these should be kept from contact with exposed parts of the metalwork as they will also remove the planned finish of the weapon as this is usually achieved by a controlled rusting process. Badly rusted areas may be lightly rubbed with fine quality steel wool soaked in a high grade light oil.

Woodwork will probably exhibit some scratches, dents or even cracks and will more often than not be 'grubby' from years of storage and use. The grime can usually be removed by wiping, as often as necessary, with cotton wool soaked in a solution of methylated spirit and water until the cotton wool comes away clean. Chequered and texture wooden surfaces will respond to similar cleaning but using a soft toothbrush in place of cotton wool. Cracks will not respond to treatment with normal wood glue as, over the years they have usually become saturated with oil. However, cleaning the edges with a substance such as 'Nitromors', will usually allow them to be bonded almost invisibly with one of the Cyano-acrylate 'superglues'. For bad breaks in the woodwork, similar prior treatment will often allow repair with a two-part epoxy resin adhesive which can be made extra hard for sanding by baking for a few minutes in an oven at about 80° centigrade. Small dents, which do not break the fibres of the wood, may often be raised by holding a moistened cloth over the dent and applying a hot domestic iron to the cloth until the dent is level with the surface of the remainder of the wood.

Once all the cleaning processes necessary for conservation have been completed, reassembly can begin. Components of the internal mechanism and areas of metal which will be covered by the woodwork should be coated with a light grease such as the virtually colourless Vaseline or one of the many similar compounds sold specifically for firearms purposes. External metalwork should also be coated with a thin oil. Once this basic groundwork has been done all that then has to be done to maintain the weapon is good condition is to follow the old military adage that it should always be 'clean, bright and lightly oiled' with just occasional repeats of the complete stripping to ensure that no new rust has appeared under the woodwork.

With no laws of security to worry about, the actual method of display chosen by the collector of deactivated weapons is completely open. On second thoughts that is not strictly true, for the collector does still have to act with responsibility. For instance, as there is no obvious outward indication of deactivation on any weapon so treated, it would be decidedly unwise to carry one openly in the streets or any public place unless it was evident that it was for a legitimate reason such as for public entertainment in something like a battle reenactment or to form part of the equipment of a restored military vehicle. These are exceptions, for most collectors will simply wish to display their weapons for the pleasure of themselves and their friends.

Below and overleaf:
While the enormous collection of the MoD Pattern Room forces different solutions.

Traditionally, weapons have been hung on walls for centuries and the majority of collectors will wish to follow this tradition. Care will need to be taken with attachment, for some military small arms are weighty — anything up to 30lb for a light machine gun — and it should be borne in mind that, although deactivated, the weapons are anything but harmless if they should fall on a foot! There are a wide variety of ready-made mounting brackets available which, if fixed to the wall properly, will be more than adequate to support the weight. Cabinets are another alternative, with the added advantage that, if totally enclosed and glass-fronted, will prevent dust settling on the weapons. This is quite a benefit for dust and gun oil will combine to form a sludge which, if left, will be quite difficult to remove.

Some weapons will not lend themselves to hanging on walls, their weight and bulk will make this impractical. In this category will be most of the medium machine guns of World War 2. However, a weapon of this type, such as the Vickers, standing in the corner of a living room, is guaranteed to break the ice at even the most frigid cocktail party and is second to none as a talking point and a conversation starter. For the more artistically inclined, there are companies around who will, for a price, take deactivated weapons and incorporate them into decorative ornamental and functional objects such as wall panels and coffee tables. In America, where DEWATS — Deactivated War Trophies — have been available for many years, some people turn rifles into standard lamps and encase pistols in acrylic blocks to form paper weights. All these are possibilities but are at the novelty end of the market and possibly not to the taste of the true collector. For the established collector of militaria, there is, of course, the option to add the weapons to a dramatic tableau of uniformed dummies of the type used for shop window displays. The options for display are really limited only by the imagination and taste of the collector and many ideas can be gained from a visit to one of the military museums who, of late, have taken great pains to display their weapons in dramatic and interesting ways.

RESEARCHING YOUR WEAPONS

The interest in collecting deactivated weapons does not necessarily end with their purchase for there is an enormous amount to be learned from the various markings that they bear. This is a very complex subject and, in order to simplify the description of the type of information that can be discovered, I will limit the explanation in this instance mainly to British service weapons.

Military weapons of all nations, particularly those with a history of long service, will be found to bear a large number of marks impressed into their metal and wood work. Many weapons will actually carry their type and model. In the case of British arms these are usually in a readily understandable form with an alphabetic name, followed by a numeric model designation, which may appear in either Roman or Arabic characters, such as 'Bren Mk II' or 'Rifle, Number 4 Mk 1'. Later minor variations to the basic models will be indicated by a further sub-designation and examples of the Bren, for instance, will be found to be marked 'Mk I (M)' with the 'M' standing for modified and announcing that the weapon was actually manufactured in Canada and differs from the standard 'Mk I' by having non-telescoping bipod legs and other minor modifications. Similarly, a submachine gun marked 'Lanchester Mk 1*' indicates that the weapon has been modified from the original pattern to allow only fully automatic fire. The weapon will also state, in some form, the name of the actual manufacturer of the weapon.

There is no standard approach to this sector of firearms marking and the amount of information varies greatly between nations. For instance, the markings of the Thompson submachine gun not only name the type and model of the weapon, it also names the manufacturer and, because it was produced and sold primarily as a commercial weapon, provides full details of all applicable patents. On the other hand, German weapons, while still providing all the same basic information, do it in a less obvious way. Model type indication will usually be a simple imprint of the official designation of the weapon such as 'P'08' for the Luger pistol or 'MP40' for the so-called 'Schmeisser' submachine gun. In most instances, the manufacturer's name only appears in a coded form such as 'byf' and 'svw' for Mauser products. Similarly, very early Walther weapons actually bore the company's 'Banner' trademark but this was later replaced on pistols by the numeric designation '480' and later still in the war by the letters 'ac'. In some cases, with German weapons, the last two figures of the year of manufacture are also stamped somewhere on the metalwork.

Also appearing on the metalwork will be various forms of official acceptance marks such as the stylised spread eagle 'Waffenamt' mark on Nazi German arms, 'ball of fire' US ordnance mark or the British arrow. These are sometimes very small, crude and indistinct and may even have been partially worn away with use. For the collector, the actual nationality of a weapon will seldom be in doubt and so these marks will be of little importance other than as a means of confirmation.

Of more potential interest will be the wide variety of small marks that appear on the weapons of some nations, particularly British. Many World War 2 British weapons were manufactured at Government arsenals, both at home and in the Commonwealth countries and each of these had its own designation which will be

Above:
The markings of military weapons clearly provide varied information. This 'Bren' gun is clearly marked as a Mk m (modified). With the additional information shown it can be deduced that the gun was made by Inglis in Canada during 1942.

found on key parts of the weapon. For instance, a Number 4 Rifle could bear the manufacturer's marks 'ROF(F)' or 'R.O.F.M', signifying production at Royal Ordnance Factories at Fazakerley or Maltby respectively, similarly the words 'Lithgow', 'Long Branch' or 'Ishapore' denote manufacture at plants in Australia, Canada and India respectively. To complicate matters, sometimes these names will have been replaced with the letters, for instance, 'RFI' or 'GRI' would identify the weapons as manufactured at Ishapore.

The Government arsenals were unable to meet the wartime demand for weapons entirely with their own resources and so they sub-contracted manufacture of components and sub-assemblies to outside commercial concerns. This led to further marks which will be found on component parts of weapons, even, in some cases, down to nuts and bolts and were only omitted where their application might weaken the part or prevent its correct function. The marks signify the manufacturer of the individual item. All authorised subcontractors were given a numeric identification code, each prefixed by a letter loosely identifying their geographic location in Britain. Quite simply 'N' equalled Northern, 'M' was the Midlands and, logically, 'S' was for the South. Therefore 'M99' would identify the source of a component as

Guest Keen & Nettlefold Ltd in Birmingham while 'S1' would be Vauxhall Motors in Bedfordshire. While, at the beginning of the war, such markings more often than not only appeared on minor components, as the conflict progressed and pressure to produce greater numbers of weapons increased, cheaper and simpler guns were developed that did not rely on dedicated ordnance machinery or on gunmaking skills for their manufacture. This meant that manufacture of components and sub-assemblies for weapons such as the Sten gun could be put out to contractors with very basic machining and metalworking facilities before being gathered together centrally for assembly by a relatively unskilled workforce. This led to a multitude of individual makers names and markings appearing on such weapons that, to the collector, can provide a very interesting picture of wartime production methods.

Once assembled, weapons were inspected for function and manufacture and again were suitably marked to indicate that this had been done. For British arms, the standard form of inspector's mark consisted of a crown surmounting a letter indicating location — such as 'E' for Enfield — beneath which was the inspector's number. The weapon was then Proof tested for safe operation and again marked. The British Proof mark was typically crossed pennants with a crown and the cypher of the reigning monarch in the top quadrant and the letter 'P' in the lower quadrant.

An interesting feature of the marking of British weapons is that often, when the weapon was passed to a unit for issue, regimental or corps marks would be impressed, commonly on the woodwork of the weapon. Such markings, were frequently obliterated with two parallel chiselled lines if the weapon was returned to stores and subsequently impressed with the marks of the new unit upon reissue. Such marks can be of great assistance to the collector in tracing the history of an individual weapon. A rifle might bear the letters YLI struck out by two parallel lines, beneath which is the inscription:

$$\frac{Y}{Fe \ \& \ Fr}$$

Above:
This Lanchester submachine gun is identified as a Mk 1*, meaning it was modified to fully automatic fire only. The letters 'SA' show the manufacturer as Sterling Armaments. The deactivation proof mark is close to the barrel while below the 'L' of Lanchester can be seen the double War Department 'Broad Arrow', signifying that it was removed from official issue. Elsewhere on the weapon, marks in arabic script show that at one stage it was in service with Egyptian forces.

Above:
Typical military proof mark that may be found on a British weapon (left). The double 'Broad Arrow' (right) will be found on weapons that have ceased to be British military property.

Above:
The letters 'ar' on the body of this 'MG42' tell us that it was built by Mauserwerke, Berlin. The quality of workmanship would indicate that it was possibly a late model. Just forward of the manufacturer's identification letters is the eagle and swastika, 'Waffenamt', official acceptance mark.

Left:
Nazi German weapons are clearly marked by model. Manufacturers are indicated by a code. In the case of this 'MP40', the letters 'fxo' identify the manufacturer as Haenel at Suhl.

Above:
The 'Waffenamt' stamp is normally only two or three millimetres across and is thus usually indistinct and poorly defined.

This would indicate that the rifle had been first issued to the King's Own Yorkshire Light Infantry and then later returned to stores before being reissued to a Yeomanry regiment, in this case the Fifeshire & Forfarshire Yeomanry. Similarly, the Officers Training Corps in some schools had their own marks, such as 'En' for Eton and 'Har' for Harrow. Stores that were not reissued and eventually became obsolete or beyond further use were marked with a pair of Broad Arrows set point to point.

A final category of marks are those used by armourers to indicate various problems or operations. In this instance an '*' marked on a barrel might indicate that rust had been found at or near the point of marking; the letters 'DP' indicate that the weapons was considered suitable only for 'Drill Purposes'.

This description has, of necessity, to be very brief and can only skim over the surface of the enormous amount of historical information available to the collector through weapons markings. Research into this area of the collection of deactivated weapons will add a totally new dimension to the hobby for the collector and will provide an immense source of satisfaction that will more than repay the effort involved.

OWEN 9M/M MK1/43 LYSAGHT P.K. AUSTRALIA

PATENTED 22/7/41 - N0115974

U.S MODEL 1928 A1
No A.O 103871

THOMPSON SUBMACHINE
CALIBRE .45 AUTOMATIC CARTRIDGE

FIRE ←——→ SAFE

FULL AUTO
SINGLE

Top left:

This Australian Owen gun has, stamped into its frames, its type, calibre and model, year of manufacture and the manufacturer's name — 'Lysaght'. On the bottom of the frame are Patent details.

Bottom left:

This Thompson submachine gun, having been sold on the commercial market before adoption for service use, not only proclaims its model and manufacturer's details but also, on the reverse side, lists all the international Patents granted.

Right:

The United States Ordnance acceptance mark takes the form of a 'Flaming Bomb'. The manufacturer, model and year or manufacture are usually clearly spelled out.

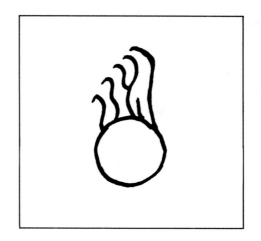

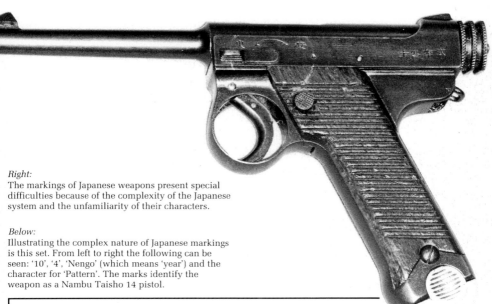

Right:

The markings of Japanese weapons present special difficulties because of the complexity of the Japanese system and the unfamiliarity of their characters.

Below:

Illustrating the complex nature of Japanese markings is this set. From left to right the following can be seen: '10', '4', 'Nengo' (which means 'year') and the character for 'Pattern'. The marks identify the weapon as a Nambu Taisho 14 pistol.

十四年式

Above:
The Japanese characters highest on the body of this machine gun reveal it to be a 6.5mm Type 96 weapon. The other characters identify the arsenal and give the date of manufacture.

Left:
The Japanese characters on this gun are, left to right: '9', '6' and 'Pattern'.

3

Rifles

The rifle has been the mainstay of armies for close on 200 years and countless generations of recruits of all nations, when introduced to the weapon, have been regaled with the prediction that it will be his 'best friend'. Be this as it may, the true importance of the position occupied by the rifle was possibly best summed up by an anonymous American officer who once said 'Keep your atom bombs, your tanks and airplanes; you will still have to have some guy with a rifle and bayonet who winkles the other bastard from his foxhole to sign the peace treaty'.

As the basic implement of war, the rifle has been less susceptible to fads and fashions than many other weapons and has generally developed more slowly. Therefore, while it is inconceivable that any major nation would have entered World War 2 with frontline aircraft or armour dating back to the end of the previous war, most armies found themselves in 1939 equipped with rifles whose designs were at least that old and many of which had actually first appeared on drawing boards in the previous century.

The majority of rifles in use in 1939 were of a similar pattern; they were usually bolt action, magazine-loaded with very similar performance characteristics and were capable of good accuracy to ranges of almost a mile, with enough energy left in the bullet at that range to inflict damage. They even looked much alike, with traditional wooden stocks that extended almost to a muzzle which was made to take some form of bayonet.

In fact the rifles of the day were very much the product of traditional methods and gunmaking skills, for although strong and simple in design, their construction was conducted with enormous care and precision; a large number of intricate machining processes were involved in their manufacture and considerable time was devoted to fitting and setting up the rifles to make them capable of tremendous accuracy. Weapons made to this kind of standard tended to last and to maintain their accuracy so, in the years leading up to World War 2, many nations had ware-

Below:
A German soldier carrying the world's first assault rifle — the 'MP44' — waves his colleagues forward. *IWM/EA48002*

Top right:
1944 in the Ardennes, an American infantryman stands guard with his 0.30in Garand 'M1' rifle. *IWM/EA51089*

Bottom right:
Troops taking part in the campaign in Burma armed with the 0.303in Lee Enfield rifles and an American 0,30in 'M1' carbine. *IWM/MH7287*

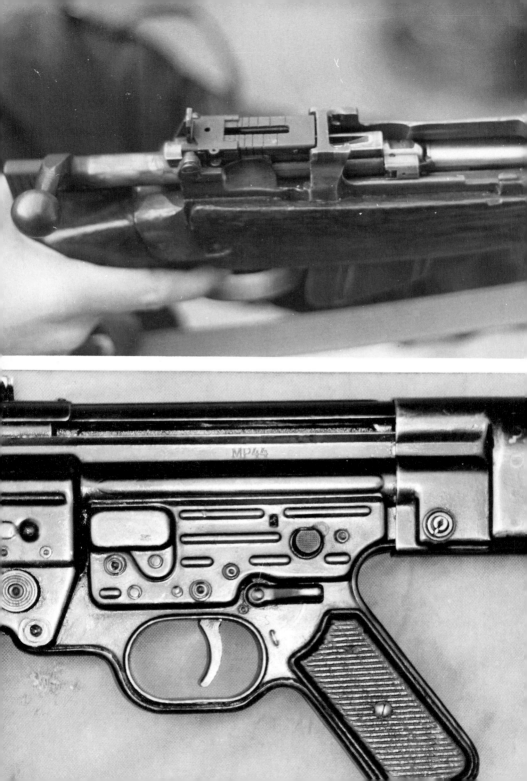

houses full of rifles left over from World War 1 that were still in excellent condition. In view of the considerable investment these weapons represented, it is hardly surprising that few nations had seen the need to devote effort to searching for a more modern replacement.

As it progressed, World War 2 was to speed dramatically the evolution of the service rifle for it was to sound the death knell of the bolt action rifle and was to usher in first the semi-automatic rifle and then, as a response to changing infantry tactics, the revolutionary new concept of the assault rifle.

The real heyday of the bolt action had begun 20 or so years before when the British Expeditionary Force had been able to demonstrate the devastating power of the magazine rifle. The ability to do this resulted from a policy of intensive weapons training instituted after the Boer War, to capitalise on the valuable lessons in marksmanship that had been learned in South Africa and to compensate for the British Government's refusal to accept the importance of equipping the army with the machine gun. Great importance was given to the ability of infantrymen to fire at least 15 accurate shots a minute with the Lee Enfield.

In 1914, when the small but extremely well trained BEF decimated the ranks of advancing German at long range, with accurate and rapid fire from their bolt action rifles, the Germans were convinced they were equipped with hundreds of machine guns and were, at least temporarily, halted. Eventually the BEF was virtually wiped out and the British Army lost its core of extensively trained marksmen. Wartime pressures did not permit the necessary time or resources to be devoted to the training of the hundreds of thousands of volunteers and conscripts to achieve the same standards of weapons usage and trench warfare eventually brought to prominence other types of weapons, requiring less extensive training and so the rifle was edged from the limelight.

The logical step forward in the development of the rifle was the use of a self-loading or semi-automatic mechanism, similar to those developed for pistols. Many designs for semi-automatic rifles had been submitted to governments from the turn of the century but, apart from a few experimental examples, had been met with little enthusiasm. The trench warfare mentality persisted and it was one of the main obstacles to acceptance. The filth and mud of battles such as the Somme were still vivid memories that dictated that any such design would have to be proved to be absolutely reliable in the most adverse conditions and in the hands of the worst soldier. Such thinking would not accept that anything but the simplest action would be capable of survival. It was also felt that semi-automatic weapons would lead to excessive and wasteful expenditure of ammunition, adequate resupply of which would be impossible.

The British were particularly firm in this belief and thus variants on the tried and tested bolt action Lee Enfield were used by the armies of the mother country and those of the British Empire throughout the war. This was not necessarily a bad thing as the venerable Lee had stood the test of time extremely well and, throughout the course of the war, it was to prove very unusual for a British soldier to acquire a German or American rifle in preference to his own. This was certainly not to be the case with other weapons such as submachine guns and pistols, which were usually exchanged at the first opportunity.

The Germans on the other hand had been experimenting with semi-automatic rifles for some years before hostilities began with designs being presented by many of the country's private arms companies. In this circumstance and with the Germans' normally excellent design expertise, it is somewhat surprising that when they did first adopt such a rifle it proved to be an abysmal failure that had to be scrapped in short order. The Germans were quick to learn from this early mistake and produce two other designs that, although produced in a climate of tremendous production problems and against enormous adverse political pressures, proved to be among the best automatic rifles ever made. The Germans were also to originate a totally new concept by rethinking the whole philosophy of the combat rifle and examining exact requirements to give the world the 'Assault Rifle'.

The Russians had also travelled a fair way along the road to developing a semi-automatic rifle but, in a repeat of the German experience, mistakenly rushed it into production early on, only to have to withdraw it in favour of an improved model, which was never able to replace the Soviet army's ancient bolt action rifle. Of all the major warring powers, the United States was unique in achieving large scale standard issue of a semi-automatic rifle to its army, with the adoption of the M1 Garand in 1936. It was largely the success of this weapon and its

Top left:
The development of the service rifle accelerated enormously during World War 2. Most nations began the war with traditional, craftsman-built, bolt-action rifles designed some 50 years before ...

Bottom left:
... As the war progressed and strain was put upon production resources, simpler and faster methods to manufacture designs had to be found. Forgings and precision machining gave way to steel pressings and stampings.

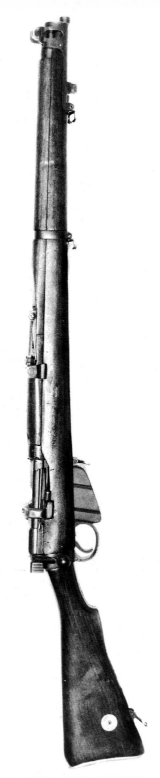

Left:
Backbone of the armies of Britain and the
Commonwealth in the early days of the war was the
Rifle No1 MkIII, a veteran of the previous war.

record of reliability in combat that led to the
acceptance of the semi-automatic rifle by other
armies.

It was towards the end of the 1930s that the
Germans went back to basics and began a
reassessment of the combat role of the rifle and
its exact requirements in service. They cut
through much of the myth, hearsay and supposi-
tions that had grown around infantry operations
and discovered that most of such actions took
place at considerably shorter ranges than had
previously been believed. From this they were
able to deduce that the conventional rifle round
was excessively large and powerful. The Germans decided that the actual maximum accurate
range required of a rifle was about 400yd, at
which range the bullet would need to retain sufficient energy to inflict damage.

Having established these optimum parameters,
the Germans were able to evolve a shortened,
lighter, 7.92mm round, based on the existing cartridge case to ease manufacture. They also specified that the best type of weapon to fire such a
round should be capable of accurate and sustained automatic fire and would therefore require
a magazine with sufficient capacity to permit
several bursts without reloading. This thinking
led to a new generation of 'Assault Rifles' that
were easier and cheaper to build, simpler to use
and which were to set the seal on rifle developments for decades to come, eventually to take
over many of the roles of both rifle and submachine gun in most armies.

BRITAIN AND COMMONWEALTH

As already outlined, the British were firmly committed to the bolt action rifle and so began World
War 2 with the rifle with which they had fin-
ished the previous one; the Short Magazine Lee
Enfield ('SMLE') or 'Smellie' as it was often affec-
tionately known. The Lee Enfield rifle had originally entered British service in 1895 at a time
when it was customary to produce two types of a
rifle; a long rifle for infantry use and a shorter
carbine model for mounted troops. The 'SMLE',

Right:
The No1 Rifle was at first supplemented and then
replaced by the simplified and easier to manufacture
Rifle No4.

accepted in 1902, introduced a new concept in military rifles, one that was shorter than the standard rifle but longer than the carbine version and hence the inclusion of the word 'Short' in its title.

The 'SMLE' was a robust rifle of fairly simple design with an underslung box magazine that held 10 rounds of rimmed 0.303in ammunition. The magazine could be loaded quickly from five round stripper clips or chargers. The 0.303in was one of the few rimmed rifle rounds in use at this stage and one of its main disadvantages was that the rifle would jam if the rims were not in the right order in the magazine. An advantage of the rifle was its method of operation. The bolt used rear locking lugs and, while this method of locking was generally considered to reduce accuracy, the 'SMLE' partly overcame this by the use of a heavy receiver. What it did allow the Lee was short bolt travel to chamber rounds, coupled with wide bolt lugs with rounded edges. When these features were added to a lightly oiled and worn rubbing surface it enabled, with a little skill and a lot of training, the bolt to be manipulated at a much faster rate than any other bolt action rifle, with less arm movement and effort.

● **RIFLE No1 MkIII*** Introduced in 1916, this rifle incorporated minor alterations the the earlier MkIII to make it simpler to build. It retained much of the outward appearance of the earlier marks with full length wooden stock extending to a blunt metal nosecap housing the foresight between winged protectors and the boss for the long 1907-pattern sword bayonet. Gone however were the sliding magazine cut-off plate and separate long range sights. Rifles of this mark were manufactured by the Royal Ordnance Factory at Enfield, BSA, the Australian arsenal at Lithgow and at Ishapore in India before finally being phased out in 1943.

● **RIFLE No4 MkI** Although an excellent rifle, the MkIII was time consuming to make and demanded large amounts of machining and hand-fitting. It was also criticised for the type and positioning of the rear sight which was an open 'U' in front of the chamber. The criticism was that it made the rifle difficult to master without a lot of practice. In 1926 an experimental simplified version of the rifle was developed and called the 'No1 MkVI'; the main alterations were incorporation of a heavier barrel, a redesigned bolt and woodwork which stopped short of the muzzle.

Left:
For jungle fighting a shorter and handier weapon was needed and this led to the Rifle No5 or 'Jungle Carbine'.

Although never adopted, this was to form the prototype of the 'No4 Rifle' which was very similar but with many minor changes to simplify production. While many of the visual characteristics of the earlier Lee Enfields were retained, the 'No4 Rifle' had a more streamlined appearance. The muzzle now protruded from the nosecap and carried the foresight assembly and bayonet boss. The rear sight was moved back to the rear of the receiver and was now an aperture sight. In fact three different types of sight may be found on this weapon, ranging from a well made adjustable leaf to a simple stamped 'L'. Weapons were manufactured at Royal Ordnance factories at Fazakerly and Maltby, by BSA, by the Savage Arms company in the USA and Long Branch Arsenal in Canada.

● **RIFLE No4 MkI*** Made in North America, this version differed mainly in that the bolt release catch, fitted behind the receiver bridge on the MkI, was eliminated and a cut-out on the bolthead used for removal.

● **RIFLE No5 MkI** By 1944, fighting in the jungle had shown a shorter version of the rifle was needed and the 'No5 Rifle' was developed. Closely based on the 'No4' which was shortened by five inches, this weapon is commonly called the 'Jungle Carbine'. The shortened barrel, much of which was exposed, reduced the muzzle velocity from the 0.303in round and made a flash hider necessary. The stock was fitted with a rubber pad to absorb the violent recoil which, coupled with greatly increased noise, made the 'No5' unpopular with its users.

GERMANY

The German attitude toward their service rifle was almost as conservative as that of the British. Thus, in the lead up to 1939, manufacturing resources were devoted almost entirely to the production of derivatives of the 1898 rifle with the Mauser action — generally rated one of the best and most successful bolt actions ever made. This success, which resulted in it being adopted, at one time or another, by most nations of the world, stemmed from the Mauser's method of locking the bolt, which made for a strong action and a high degree of inherent accuracy. By using lugs at the front of the bolt, which locked into recesses in an extension of the barrel, stress was eliminated from the bolt and the walls of the

Right:
Mainstay of the German army was the Mauser-designed 'Kar 98k' carbine.

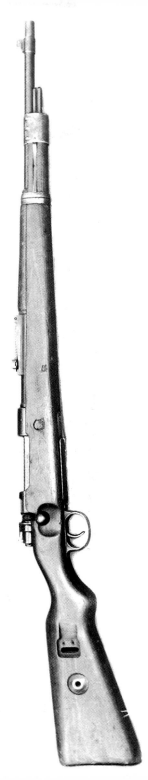

receiver, allowing them to be made relatively light. Penalties were that bolt travel was increased by the extra length added to the bolt by the forward lugs and it therefore could not be worked as quickly as for instance the rear locking Enfield action; also the locking recesses in the barrel extension were difficult to keep clean and free from debris.

Throughout the war, the Germans were never able to produce enough rifles and used a large variety of weapons, adopting those of captured countries as well as specimens captured from the Allies and there are consequently relatively few types built by the Germans and which can be considered as standard issue.

● **Mauser 7.92mm Kar98k** This was the standard issue rifle and with it the Germans had followed the same course as the British by shortening the standard rifle by six inches to produce the 'Karabiner' or 'Kar98'. The rifle was accepted into service in 1935 and in length and weight was very similar to the 'SMLE' although the 'Kar98k' had much slimmer looking woodwork finishing short of the muzzle. The bolt handle was bent downwards and a hollow carved in the stock to house the head. The foresight was an exposed post on the muzzle and the rear sight placed well ahead of the chamber. This produced a short sight radius, detracting from the accuracy of the weapon. The bayonet was carried by a bar below the muzzle. In models made in 1944 and 1945, the bayonet bar was removed and, as supplies of good timber became scarce in Germany, the rifle was fitted with laminated wooden stocks. The majority of 'Kar98' rifles were assembled by Mauserwerke AG although many component parts were subcontracted. Examples were also built at the captured FN factory at Liege in Belgium and similarly at the Czechoslovakian Brno factory. The quality of the Mauser rifle was generally maintained at a high level throughout the war although, towards the end, rather suspect cast receivers were used to overcome material shortages and to speed production.

Some interesting accessories were provided for the 'Kar98k', including an enlarged trigger guard to permit the use of gloves and a special winter trigger mechanism in the form of a sheet metal cover for the trigger guard and an external lever to make it possible to actuate the trigger, even with the thickest gloves.

● **Gewehr 33/40** A version of the Czech Model 33 carbine made by the Brno factory in Czechoslovakia during occupation. It is the shortest barrelled version of the Mauser adopted by the Germans and, because of its light weight and short barrel, was subject to severe blast and recoil. The receiver is cut to lighten the weight and the butt plate is extended on the right side of the stock. Its use was mainly by mountain and paratroop regiments, for whom a version with a hinged stock was produced 'Gewehr 41(W)'. In 1937 the

Germans began to look seriously at the possibility of adopting an automatic rifle. By 1941 two different designs were ready for combat trials; the 'Model 41(M)' and 'Model 41(W)', made respectively by Mauser and Walther. Both weapons used a variation of the Bang system of operation where a muzzle cone traps gas and causes it to act against a piston to work the action. The Mauser version was unsuccessful and was withdrawn in favour of the Walther design. The operating rod of the '41(W)' was above the barrel and the bolt used a locking system in which a pair of flaps were pushed outward to lock receiver and bolt while the firing pin moved forward. Several thousand of these rifles were built, mainly for use on the Eastern Front, where they were lost or discarded, for the rifle was never popular. It is a rare rifle in the West although the opening of the Eastern Bloc may see more examples becoming available.

● **Gewehr 43** Combat experience with the 'G41(W)' quickly demonstrated areas of potential improvement. The muzzle cone operating system, which had been shown prone to fouling was changed in favour of a more conventional piston type, with gas tapped through a port in the barrel. The whole weapon was lighter, better balanced and easier to build then its predecessor. It was also unusual, for a military rifle, in that a dovetail was machined on to the receiver for a telescopic sniper sight. Again the weapon was most extensively used on the Eastern Front. Quality was high although, towards the end of the war, external finish suffered and the high quality wooden stock of the early models was replaced with laminated wood or even crude phenolic resin compounds. After the war the Czechs retained the 'G43' as their standard sniping rifle.The 'G43' was manufactured by Carl Walther Waffenfabrik, Zella-Mehlis, Berliner-Lubecker Maschinenfabrik AG, Lubeck and Gustloffwerke, Suhl.

● **Maschinenpistole 43/Sturmgewehr 44** Quite probably the most significant small arms development to come out of Nazi Germany was the introduction of a new concept in service weapons — the 'Assault Rifle'. Having analysed basic service rifle requirements and arrived at the decision that the traditional rifle round was unnecessarily large and powerful for most com-

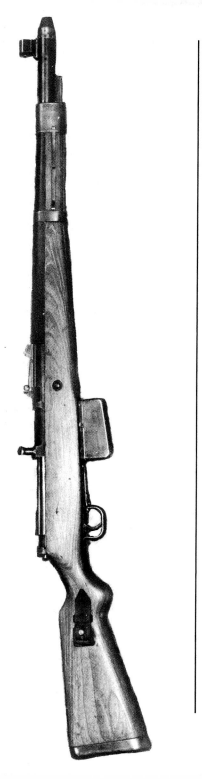

bat situations, the Germans arrived at the 'inter-mediate' 7.92mm Kurz (short) round. While the same basic calibre as standard 7.92mm x 57mm rifle round, the cartridge case was only 33mm long and contained 24.6 grains of propellant against the 45-50 grains in the previous service ammunition.

Having decided parameters for the new ammu-nition, attention was given to the type of rifle needed to fire it. The requirement was for a weapon that would firstly have some of the attributes of the submachine gun in that it would be light, handy and able to fire fully automatic without excessive recoil; second it would act as a semi-automatic rifle for normal infantry opera-tions, accurate to about 400yd and with 'punch' at that range; lastly it would have to take on the light support role and deliver automatic fire, rea-sonably accurately out to 400yd or so. In addi-tion, it needed sufficient magazine capacity to enable the user to fire several bursts without reloading.

Contracts were placed with Carl Walther and Haenel for the development of suitable weapons both of which were ready in limited quantities for combat testing in 1942. The Haenel version was eventually chosen for quantity production and entered service as the 'Maschinen Pistole 43'. Interestingly, the misnaming of the weapon as a machine pistol seems to have been because Hitler, based on his experiences as a World War 1 infantryman, initially rejected the weapon as a rifle as it did not have the long range capability he felt necessary. Classifying it as a machine pis-tol was a back door way of continuing produc-tion.

Considerable attention was given to making the weapon as simple as possible to manufacture and, early in the development process, a firm of specialist steel pressing specialists were involved in the design. As a result, the 'MP43' set new standards of rifle manufacture. Gas operated, with a piston housed in a tube above the barrel, the 'MP43' used a straight-line recoil path into the stock which greatly increased controllability in automatic fire. A centralised production sys-tem was instituted, with sub-contractors all over Germany manufacturing parts for the weapon, many of which were stampings or pressings which were riveted or welded together. Early models were able to take a clamp-on grenade launcher although this was quickly changed to a screw on type. For no apparent reason, other than the changing year, the weapon was reclassi-

Left:
Towards the end of the war, the Germans introduced a new class of weapon — the 'Assault Rifle' — which was designated 'MP44'.

fied the 'MP44' and, shortly after, took on the more accurate description, 'Sturmgewehr' or 'assault rifle 44'. Examples were manufactured by C. G. Haenel, Suhl; Erfurter Maschinenfabrik B. Giepel, Erfurt and by Mauserwerke AG, Oberndorf am Neckar. An interesting variant the, 'Maschinenpistole 44 mit Krummlauf' was manufactured by Haenel and by Rheinmetall-Borsig in Dusseldorf. This was an 'MP44' to which a curved barrel unit and a mirror sight were attached. Several unprovable theories are put forward for this unlikely device but the most widely accepted is that it was for shooting around corners without exposing the shooter. Whatever the true explanation, the Krummlauf device allowed bullets to be fired at angles of 30° or 90° according to model.

● **Fallschirmjagergewehr 42** Another outstanding German automatic rifle design, although not truly an infantry weapon, was the 'FG42'. Originated for the paratroops, who came under Luftwaffe control, the rifle was designed to have all the characteristics of an assault rifle but using full powered rifle ammunition. The 'FG42' was gas operated but had several novel features. For instance, it fired from the open bolt to aid cooling when set at 'automatic' but from the closed breech, improving accuracy, when set for single shot. The weapon, which weighed less than 10lb, had the straight line 'bullpup' design that is fashionable now and was fitted with a bipod and integral bayonet. Ammunition was fed from a 20-round box magazine on the left side of the receiver. It is likely that no more than 8,000 were ever built as the weapon was expensive and time consuming to manufacture and the paratroops faded from importance. Two types were manufactured. Early models had a stamped steel stock and a sharply raked pistol grip; later versions a wooden stock and a vertical pistol grip. The weapon was manufactured by Heinrich Waffenfabrik, Suhl and by Rheinmetall.

ITALY

The Italians started the war with one of the oldest and worst bolt action rifles of any of the fighting nations. The basic rifle was the 'Mannlicher-Parravicino-Carcano Model 91'. The convoluted name describes a basic Mauser bolt action with forward locking lugs, to which is added a Mannlicher magazine and a safety catch from

Right:
Italy was another nation wedded to a bolt-action rifle, the complex and unsatisfactory 'Mannlicher Carcano' which was manufactured in a variety of forms. The model illustrated here in the 'M38' carbine.

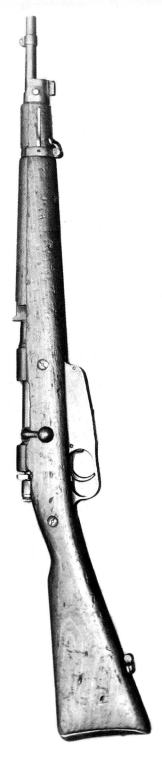

Carcano. Parravicino was the officer in charge of its original development. The weapon fired a 6.5mm round with a light, round-nosed bullet that allowed low chamber pressures and thus light recoil but which also quickly lost its velocity and effectiveness with range. With such a relatively low powered round, the Italians were able to trim weight from the rifle by machining unnecessary metal from the receiver, a fact which has led to the belief that the rifle was weak and dangerous.

The Italians compounded their problems by deciding, in 1938, to uprate performance by changing calibre. They chose 7.35mm, which meant evolving a new cartridge case and a new barrel, although the intention was to keep chamber pressure at the original level by employing a lighter, pointed bullet with an aluminium nose-cap. When war came, money ran out and most of those rifles that had been converted were converted back to 6.5mm. The Italians have always borne the brunt of much joking about the contribution of their army to the war. In many ways this is unfair, for their troops were poorly trained and even more poorly equipped. With training of the type enjoyed in other armies and equipment even remotely similar in quality and performance it might have been a different story.

● **Mannlicher Carcano Modello 91** The basic rifle was developed at the Turin Arsenal in 1890 and uses the Mannlicher clip system. Magazines of rifles such as the Enfield and Mauser are loaded from 'chargers' — disposable metal clips which hold the bases of the cartridges and from which the cartridges are stripped into the magazine by thumb pressure. In the Mannlicher system, the clip remains attached to the rounds and forms a part of the magazine system. A follower arm in the magazine housing forces rounds out of the clip for the bolt to push into the chamber and, when the last round has been chambered, the empty clip falls out through the bottom of the housing. Unfortunately, the open magazine housing also made an excellent scoop for mud and debris. The rear sight was a simple 'V' notch.

● **Model 1891/24** In line with other nations, Italy decided to eliminate the separate rifle and carbine. The barrel of the 1891 rifle was shortened, the bolt handle bent down and the sights improved. This became the standard army rifle although there were numerous rifle and carbine variants, some with attached folding bayonets.

● **M1938** A slight redesign was forced by the introduction of the 7.35mm cartridge which was too powerful to be fired comfortably from the short barrel of the 'M1891/24'. A rifle with a 21in barrel was developed. With the return to 6.5mm calibre, quantities of this rifle were reconverted. A fixed rear sight set at 300m is a feature of this weapon. An 'M1938', with a cheap telescopic sight, is said to have been used to assassinate President Kennedy in Dallas.

50

At the outbreak of war, Japan, like Italy was caught in the middle of a calibre change and so there were two types of cartridge used by the Imperial Japanese Army — 6.5mm and 7.7mm. The 6.5mm was a semi-rimmed round, dating from around 1897 which shared most of the failings of the Italian 6.5mm although it was, in many ways, different. It was replaced in 1939 by a rimless 7.7mm calibre, a more powerful cartridge but time and production problems did not allow the Japanese to get this into general issue with the Imperial Army.

It should be noted that the Japanese used a different form of designation for their weapons. Whereas most Western weapons bear their date of introduction, according to the Gregorian calendar, somewhere in their designation (eg 'Kar98K', 'MP40', etc), the Japanese dated their weapons from the year of the reign of the Emperor in which they were introduced. For instance, the 'Meiji 30' rifle , was introduced in the 30th year of the reign of the Emperor Meiji, which was 1897 in our calendar.

In all political and industrial respects, the Japanese had remained virtually a medieval state until the latter half of the 19th century, after which their transition into a modern nation was comparatively rapid. The country had no real practical experience of contemporary weapon technology and so it was not too surprising that when, around the turn of the century, a commission, headed by a Col Arisaka, were tasked to investigate and recommend improvements to the equipment of the army, that they chose to adopt versions of the Mauser rifle and which bore Co Arisaka's name.

● **MEIJI 38** This rifle was introduced in 1905 and differs from the 1898 Mauser in the arrangement of the safety catch, which takes the form of a large knob on the rear of the bolt. Also, unlike the normal Mauser action, the mechanism was cocked on closing the bolt. The rifle was also fitted with a metal cover which reciprocated with the bolt to keep out dust and rain. However, the operation of this proved to be too noisy in close-quarter jungle fighting and was almost invariably discarded. It is therefore unusual to find examples of the rifle with this cover still in place. The bolt handle of the rifle is not bent down and stands at right angles to the action. Ammunition is fed from an integral five round magazine. A distinctive feature of all Japanese rifles of the period is that the butt is made from two pieces of wood with the lower half of the butt and pistol grip pinned and glued to the remainder. The bayonet for this rifle was a long sword-type.

Left:
Standard service rifle of the Imperial Japanese Army was the 'Arisaka', which appeared in a variety of styles and a confusion of calibres. This is a 'Type 38' long rifle. A characteristic of Japanese rifles is the two-piece stock, jointed horizontally.

Above:
The 'Type 38' rifle was issued with a sliding sheet metal cover for the action. However, this rattled and was often discarded.

Above:
Many 'Arisaka' rifles carry a representation of the national flower — the Chrysanthemum — on the bolt.

The 'Type 38' was produced in enormous numbers during the war. Some estimates which, through lack of documentation, cannot be substantiated, put the figure as high as 10 million. The quality of the Arisaka rifle is general high although after 1943 the standard began to drop until the final examples were plain dangerous to the user. Carbine and take-down versions were also produced.

● **TYPE 99** This model was develoned to use the 7.7mm round and was very similar to its predecessor but with a folding wire monopod under the barrel to provide support. The value of this would seem to have been minimal as it was extremely fragile. Also of limited value would have been the anti-aircraft sight which was fitted as standard equipment. This took the form of a pair of arms hinged to each side of the rear sight and intended to indicate the amount of lead needed to tackle aircraft crossing in front of the shooter. The 'Type 99' was made in two lengths — a standard version and a 'short' model which was in line with the contemporary European thinking. All Arisaka rifles were manufactured at the Japanese State Arsenals.

SOVIET UNION

The armies of the Soviet Union and of Italy share the dubious distinction of having entered World War 2 with the oldest rifles of any of the major powers. The Mosin-Nagant was originated for the Imperial Army in 1891 as a collaboration with the Belgian gunsmith Emile Nagant and

Right:
The Soviet 'Moisin Nagant 1891' rifle.

Far right:
The 'M1938' carbine was in reality just a scaled-down version of the rifle but without the ability to accept a bayonet.

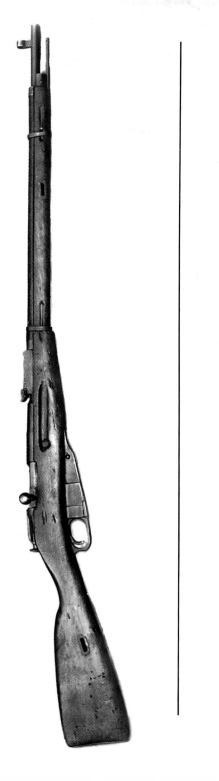

53

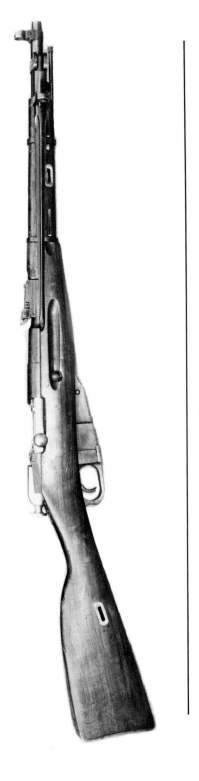

54

S. I. Mosin, a Russian artillery officer who became director of the Imperial Armoury at Tula.

The resulting Mosin-Nagant was a fairly conventional Mosin-designed bolt action allied to a magazine designed by Nagant. The bolt is a three-piece design that seems to owe its over-complexity to the need not to infringe contemporary patents rather than to any practical considerations. It locks with two lugs at the front of the bolt which turn into recesses in the breech. Another lug locks the rear of the bolt. The Nagant magazine has an unusual form of latch control which holds down the lower rounds in the magazine to leave the round to be chambered free of spring pressure during loading. With the bolt closed, the latch allows the next round to rise to the underside of the bolt and then moves in once again to hold the next rounds down while loading is repeated. Prior to the Revolution, when metric measures were introduced, the calibre was measured in an ancient Russian unit — the 'Line'. The sights were also calibrated in another archaic measure — the 'Arshin' which was based upon the length of the human pace.

● **M1891/30M** This is a basic modernisation of the original 1891 rifle design with improved, metric sights. It equipped a large part of the Soviet army and, with a carbine version, remained a standard infantry weapon until 1950. A sniper version was also produced with a turned-down bolt handle. The Mosin Nagant was a well made and reliable weapon that was accurate and easy to maintain in the field.

● **Carbine M1944** A shortened version of the standard rifle designed primarily for mounted troops but issued much more widely and, in large part, supplanted the full length rifle. The carbine featured a permanently fixed folding bayonet.

● **Tokarev SVT-40** The Russians, together with the Americans, were the only major powers to make an early commitment to the introduction of an automatic or self-loading rifle for its troops. Designed by Feodor Vassilevitch Tokarev, the 'SVT-40' was gas operated, with gas being bled

Far left:
A late variant was the 1944 carbine, which had a cruciform folding bayonet permanently attached.

Left:
The Soviet Union's first self-loading rifle was the 'SVT40'. A virtually identical selective-fire model — the 'AVT40' — was also introduced. It is identified by a modification to the safety catch, permitting automatic fire.

Right:
An early version of the 'SVT40'.

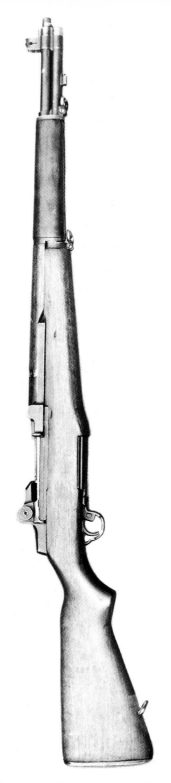

off from close to the muzzle to drive a piston, which, in turn, operated a rod to strike the face of a bolt carrier to drive it to the rear against spring pressure. The bolt itself was locked by cams until chamber pressure dropped and the return spring was in the bolt carrier, with a separate spring to return the operating rod and gas piston. An unusual feature was flutes cut into the chamber to allow gas to leak around the case so that it 'floated' in the chamber and was thus prevented from sticking to the walls. The rifle was issued in significant numbers. A variant, the selective fire 'AVT-40' was less widely used, this is believed to be because the mechanism proved unreliable under sustained automatic fire.

UNITED STATES

The United States of America was the only major power at the outbreak of war to have an army almost completely equipped with an effective self-loading rifle and it remained the only army so equipped throughout the conflict. The search for such a weapon had begun in 1929 with a series of competitive trials at the Aberdeen Proving Grounds which led eventually to the adoption of a design by John C. Garand, designer at the Springfield Armoury. His rifle was gas operated and chambered for a new .276in cartridge. However, Gen Douglas MacArthur, then Chief of Staff, strongly opposed the adoption of the .276in cartridge on the grounds that the country held massive stocks of and had tremendous investment in the manufacture of the 0.30in-'06 and this could not be wasted. John Garand went back to his drawing board and modified the design and, in 1936 his rifle was officially adopted as 'US Rifle, cal 0.30in M18' but became world renowned as the 'Garand'.

With only minor alterations, the 'Garand' served as the main weapon of the American armed forces throughout both World War 2 and the Korean War, building a reputation for reliability that prompted Gen Patton to call it 'The best battle implement ever devised'. It was the proven reliability of the 'Garand' that eventually prompted many other nations to follow suit and adopt the self loading rifle.

● **Rifle 0.30in Cal M1** The 'Garand' was a simple and very strong rifle. Gas was tapped from near the muzzle into a cylinder immediately below the barrel where it drives a long stroke piston towards the rear. Acting on a cam, this action

Left:
The United States was the only major power to begin World War 2 with an army equipped with a self-loading rifle. The Garand 0.30in-'06 'M1' was the basic infantry weapon.

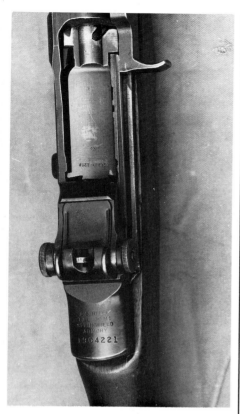

Above:
The solid, traditionally-built action of the 'M1' rifle.

rotates the bolt to open it and, at the same time, cocks the firing hammer. The return spring is also contained in the cylinder and this allows the body of the action to remain fairly short. The bolt is locked by two forward lugs engaging in recesses just behind the breech. In appearance, the 'Garand' is an attractive weapon with a wooden stock that extends half way along the barrel. A wooden handguard protects the top and remaining third, with the final few inches of barrel and gas cylinder left exposed.

Right:
The 0.30in 'M1' carbine was introduced to replace the pistol for NCOs, company grade officers and special forces. A selective-fire 'M2' version and an 'M3' with receiver grooved to take an infra-red nightsight were also produced.

The 'Garand' was almost universally liked and respected by its users. The one area of criticism was the magazine and its method of loading. Eight rounds of ammunition are held in a clip inside an integral magazine. The rifle had to be loaded by pressing a complete clip of eight rounds into the magazine where it remained until empty, whereupon it was ejected upwards. This method held a number of minor disadvantages. For instance, the rifle could only be loaded with a complete clip and could not therefore be 'topped up' with loose rounds. The clip also made a distinctive 'ping' when ejected and clattered on landing on hard surfaces, advising all and sundry that the rifle was empty.

There were few variants of the basic 'Garand'. Two sniper's versions, the 'M1C' and 'M1D' were produced, using a Griffin & Howe telescope mount and either a Weaver 330 or Lyman 'Alaskan' telescopic sight. In some cases a leather cheek pad was also strapped to the butt. A conical flash hider could also be clipped over the muzzle and attached to the bayonet mount. A very small number of shortened 'M1s', officially designated 'T26' were produced late in 1945. This very rare variant is often referred to as the 'Tankers' model'.

The 'Garand' was manufactured during the war mainly by the Springfield Armoury and the Winchester Repeating Arms Co, with extra manufacturing capacity supplied by International Harvester and Harrington & Richardson.

● **US Carbine Cal 0.30in M1** In 1940 the American forces began to look for a light rifle to replace the pistol and standard rifle for arming troops such as machine gunners and drivers, whose prime function was not to use a rifle but who, in emergency, might need a more potent weapon than the pistol. A specification was drafted based upon a new straight sided, rimless round of ammunition derived from a sporting cartridge, the 0.32in Winchester Automatic. Eleven designs for weapons were submitted for consideration before eventually, one from Winchester Repeating Arms Co, using a modification of the 'Garand' rotating bolt but with a short stroke gas piston was adopted.

The gas action used a captive piston, which travelled only one third of an inch. This struck an operating slide, transferring sufficient energy to open the bolt against the return spring and to cock the hammer before the spring returned the bolt, chambering a new round. The whole mechanism was wrapped in a pretty (if one can actually use that adjective of a weapon) wooden three

Right:
The 'M1A1' carbine replaced the wooden stock with a folding metal version.

quarter stock which gave it an almost sporting appearance. Well over 6 million carbines were manufactured and distributed throughout the American Army and Marine Corps during the war, making it the American's most widely produced small arm. The introduction of the 'M1' carbine was in fact a retrograde step in firearms development for, at the time of its introduction, in many other armies, the submachine gun had been adopted for the roles laid down for the 'M1'. Perhaps it was felt that as the heavy, complicated and expensive to manufacture 1928 Thompson was the US Army's issue submachine gun, the introduction of a new type of weapon was justified.

The true effectiveness of the 'M1' carbine has always been open to debate for, although light and easy to use, it was neither 'fish nor fowl' having neither the power of the rifle nor the handiness of the pistol or submachine gun. Also, its ammunition offered neither accuracy or stopping power and, for these reasons the 'M1' was one of the most often discarded of all American weapons, particularly in its selective fire 'M2' variant. Ammunition was fed from 10, 15 or 20-round detachable box magazines and the 'M1' could be fitted with a knife bayonet. Early models were made with an 'L' type flip-over aperture rear sight which was changed in later production by a ramp-mounted aperture sight that was adjustable for windage.

● **US Carbine Cal 0.30in M1A1** Operationally identical to the 'M1' but fitted with a pistol grip and a folding metal skeleton stock intended primarily for airborne forces.

● **Carbine Cal 0.30in M2** A selective fire version, introduced in 1944 and outwardly the same as the 'M1' but with a change lever on the left of the receiver. A special 30-round magazine was developed for this model which had a cyclic rate of fire of 750 rounds a minute.

● **Carbine Cal .030in M3** A 1945 development of the 'M2' in which the open sights were removed to allow fitment of a large infra-red sniperscope.

The 'M1' carbine was manufactured by a large number of companies: Winchester Remington Arms Co., New Haven Connecticut; The Inland Manufacturing and the Saginaw Steering Gear Divisions of General Motors, Grand Rapids Michigan; Underwood-Elliot-Fisher, Hartford Connecticut; Rochester Defense Corp, and National Postal Meter Corp, Rochester, New York; Quality Hardware Corp and the Rock-Ola Corp, Chicago, Illinois; Standard Products Co, Port Clinton, Ohio; and IBM, Poughkeepsie, New York.

Right:
With stock folded the M1A1 carbine was ideal for airborne forces or troops, such as tankers operating in confined spaces.

Rifle Technical Data

Britain & Commonwealth

Type	Calibre	Overall length	Barrel length	Weight	Feed	Sights	Muzzle velocity
Rifle No1 MkIII*	0.303in	44.5in	25.19in	9.25lb	10-round detachable box	(Front) Blade in protecting ears (Rear) Tangent leaf	2.440ft/sec
Rifle No4 Mkl	0.303in	44.5in	25.2in	8.8lb	10-round detachable box	(Front) Blade in protecting ears (Rear) Vertical leaf or 'L' type battle sight	2.440ft/sec
Rifle No5 Mkl	0.303in	39.5in	18.in	7.15lb	10-round detachable box	(Front) Blade in protecting ears (Rear) Vertical leaf with aperture battle sight	2.400ft/sec

Germany

Type	Calibre	Operation	Overall length	Barrel length	Weight	Feed	Sights	Cyclic rate	Muzzle velocity
Kar98k	7.92mm		43.6in	23.6in	8.6lb	5-round fixed magazine	(Front) Barley corn (Rear) Tangent leaf	M2.476ft/sec	
Gew 33/40	7.92 x 57mm		39.1in	19.29in	7.9lb	5-round fixed magazine	(Front) Barley corn (Rear) Tangent leaf		2.400ft/sec
GEW 41(W)	7.92 x 57mm	Gas. semi-automatic only	44.24in	21.5in	11.08lb	10-round fixed magazine	(Front) Barley corn (Rear) Tangent leaf		2.550ft/sec
GEW 43	7.92 x 57mm	Gas. semi-automatic only	44in	21.62in	9.5lb	10-round detachable box	(Front) Barley corn (Rear) Tangent leaf		2.550ft/sec
FG42	7.92 x 57mm	Gas. selective fire	37in	21.5in	9.93lb	20-round detachable box	(Front) Folding barley corn (Rear) Folding aperture	750-800rpm	2.500ft/sec
MP43/StG44	7.92mm kurz	Gas. selective fire	37in	16.5"	11.5lb	30-round detachable box	(Front) Hooded barley corn (Rear) Tangent	500rpm	2.132ft/sec

Italy

Type	Calibre	Overall length	Barrel length	Weight	Feed	Sights	Muzzle velocity	Attachments
1891/24/1938 Rifle*	6.5mm/7.35mm	36.2in/40.2in	17.7in/20.9in	6.9lb/7.5lb	6-round clip/6-round clip	(Front) Barley corn/Barley corn (Rear) Tangent/Fixed	2.297ft/sec/2.482ft/sec	Folding bayonet
1938 Carbine*	7.35mm	36.2in	17.7in	6.5lb/6.8lb	6-round clip	(Front) Barley corn (Rear) Fixed	2.400ft/sec	
1938TS Carbine*	7.35mm	36.2in	17.7in	6.8lb	6-round clip	(Front) Barley corn (Rear) Fixed	2.400ft/sec	Folding bayonet

* = Versions were also produced in 6.5mm with only minor variations in weight and muzzle velocity.

Japan

Type	Calibre	Overall length	Barrel length	Weight	Feed	Sights	Muzzle velocity
Meiji 38 Rifle	6.5mm	50.2in	31.4in	9.25lb	5-round box	(Front) Barley corn with protecting ears (Rear) Leaf	2.400ft/sec
Meiji 38 Carbine	6.5mm	34.2in	19.9in	7.3lb	5-round box	(Front) Barley corn with protecting ears (Rear) Leaf	2.300ft/sec
Type 99 Long Rifle	7.7mm	50in	31.4in	9.1lb	5-round box	(Front) Barley corn (Rear) Leaf	2.390ft/sec
Type 99 Short Rifle	7.7mm	43.9in	25.8in	8.6lb	5-round box	(Front) Barley corn with protecting ears (Rear) Leaf	2.360ft/sec

Soviet Union

Type	Calibre	Length	Barrel length	Weight	Feed	Sights	Muzzle velocity
M1891/30 Rifle	7.62mm	48.5in	28.7in	8.7lb	5-round box	(Front) Hooded Post (Rear) Leaf	2.660ft/sec
1944 Carbine	7.62mm	40in	20.4in	8.9lb	5-round box	(Front) Hooded Post (Rear) Leaf	2.514ft/sec
Tokarev SVT-40	7.62mm	48.1in	24.6in	8.59lb	10-round box	(Front) Hooded Post (Rear) Tangent	2.519ft/sec

United States

Type	Calibre	Length	Barrel length	Weight	Feed	Sights	Muzzle velocity	Cyclic rate
US Rifle M1 'Garand'	0.30in — '06	43.0in	24in	9.5lb	8-round clip loaded box	(Front) Post with protecting ears (Rear) Adjustable aperture	2.800ft/sec	
Carbine Cal 0.30in M1	0.30in	36in	18in	5lb	15-round box	(Front) Hooded Post (Rear) Flip over 'L' peep	1.950ft/sec	
Carbine Cal 0.30" M2	0.30in	36.65in	18in	5.45lb	15/30-round box	(Front) Hooded Post (Rear) Aperture	1.950ft/sec	750 rounds/min

4

Pistols

Of all firearms, the handgun or pistol probably inspires the imagination more than any other and has, in the process, gathered around it a powerful mystique and glamour. It also has the most varied and interesting history and also one of the longest, some of the earliest applications of gunpowder and some of the most significant subsequent developments having appeared first in hand weapons. Throughout history, the pistol has often been the weapon of last resort and has thus been a very personal piece of equipment and, in some military circles, a private purchase by officers rather than a general issue item to other than specialist troops. As is so often the case, the mystique hides a far different reality for, militarily, the pistol has only limited application. The true effectiveness of the handgun as a weapon of warfare was perhaps put into context by one British General who recorded that of the 30 men he had seen wounded by pistol fire during war, 29 had been his own men mishandling their guns, the other was an enemy he himself shot.

Traditionally, pistols have been issued by the military for the personal defence of troops armed with weapons other than rifles and as a symbol of status for officers. Its main drawback as a general issue weapon is that, while the pistol and its ammunition are generally inherently accurate, a large commitment of time has to be made to training before it may be used effectively, and this training absorbs resources which may be better devoted to more potent weapons. For these reasons, most armies issued pistols only to officers and to specialist troops, such as commandos, who were able to devote the necessary time to training and who would often be involved in close quarter combat. Some armies, that of the Soviet Union for instance, almost totally forsook the pistol in favour of weapons such as the submachine gun which were simpler and much cheaper to make; easier to train, plus being far more effective. Pistols are also intricate pieces of equipment requiring precision construction and fitting and are therefore expensive in terms of time and material to manufacture.

From the beginning of this century, there were two major types of pistol — revolvers and semi-automatics — available. Throughout the 19th century, the revolver had held sway as the military sidearm but towards the end of the century, Hiram Maxim, with his machine gun, had demonstrated that it was possible to harness the power of the cartridge to operate the mechanism of a machine gun.

Almost immediately, weapons designers turned to the application of variations of this principle to handguns and with this development began an argument of the comparative merits of

Above:
By the beginning of World War 2, most nations had at least toyed with semi-automatic pistols for service use. The nearest Britain came to success was the odd ball Webley Fosbery semi-automatic revolver.

Left:
Early in the war, men of the Waffen SS Regiment 'Germania' relax. The soldier on the left clutches a 'P08' pistol while his colleague second right has the standard Mauser 'Kar 98' rifle. *IWM/MH225*

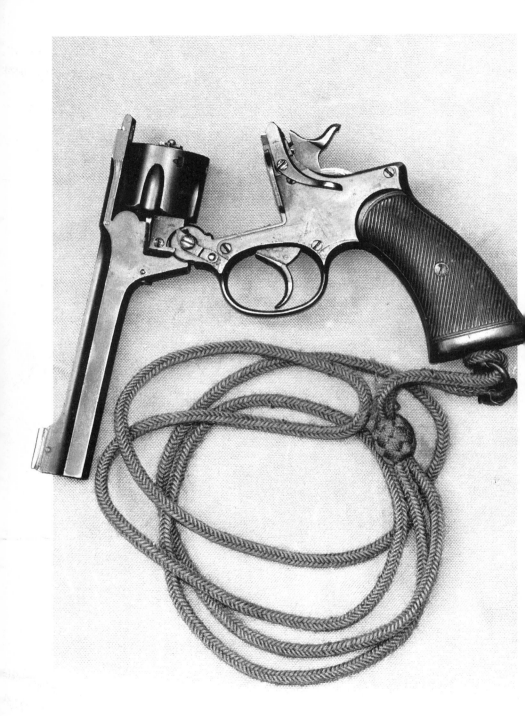

revolvers versus semi-automatics as a combat weapon that has raged almost continually since. Certainly, by the outbreak of World War 2 this question had not been resolved and both types were to be found in use in most armies. The argument is briefly that revolvers are generally simpler, stronger and more reliable while semi-automatics are lighter, have greater ammunition carrying capacity and fire faster. However, this type of weapon was considered less safe in inexperienced hands as, after each shot it is automatically reloaded and left in a cocked condition, ready to fire at the next squeeze of the trigger.

It is worth, at this point, clarifying the definition of the term 'automatic' pistol. While this term is generally used to describe the normal magazine-fed pistol, the more accurate designation is semi-automatic or self-loading. In operation, the weapon is usually loaded by manually

Left:
The British did however move from the bulky break-action Webley revolver ...

Below:
... to the more compact and modern self-loading pistol during the war.

pulling back the breech block and then allowing it to run forward, chambering a round and leaving the firing mechanism in the cocked position. When the trigger is squeezed, the round is fired and part of the energy this develops is used to recycle the mechanism, ejecting the spent case, chambering a fresh round and recocking the mechanism.

A disconnector in the trigger mechanism prevents the firing mechanism tripping until the trigger is released and squeezed again. The firing cycle is therefore semi-automatic, requiring the shooter to make a conscious decision and make a physical effort to fire a subsequent shot. In a true automatic pistol, while the weapon is initially manually charged, one squeeze of the trigger will cause the pistol to continue to fire until either the magazine is empty or the trigger released. Most weapons of this type incorporate a change lever to permit semi-automatic operation when required. The true automatic pistol was really a precursor of the submachine gun and has largely been replaced by it.

The first designer to achieve real practical success with a semi-automatic pistol was German-born Hugo Borchardt, who put his pistol into production in 1893. Although it was technically very advanced, using a variation of the Maxim toggle-link, the Borchardt pistol was clumsy and delicate and was thus not a commercial success until

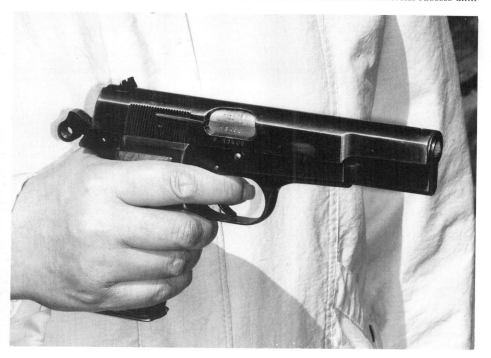

Below:
Arguably the best-known is the German P'08 or
'Luger', with its distinctive but sensitive 'toggle-link'
locking system.

another German, Georg Luger, simplified the mechanism and improved it into the 'Pistole 08', popularly known as 'The Luger'.

Most commercially successful of the very early European semi-automatic pistols was however the 'Broomhandle Mauser'. It was so successful in fact that it practically eliminated the revolver as a service weapon with the majority of European armies forever more. Its main advantage was that, because it was superbly built, it was absolutely reliable in operation. Although it only fired an 86 grain bullet, it did so at an almost incredible (for the time) 1,400ft/sec. Superbly accurate, a feature enhanced by the ability to clip its wooden holster on to the butt as a shoulder stock, its standard 10-round magazine, loaded rapidly from stripper clips, permitted phenomenal rates of sustained fire.

Although a sound and reliable weapon, the Mauser had an appalling grip and was extremely cumbersome, more comfortably used as a light carbine than as a pistol. It did remain in service with some armies into World War 2, it, was soon to make way for less complex and more effective designs from a man who was to prove to be one of the most prolific and ingenious of weapons designers, John Browning. Between 1896 and his death in 1926, Browning was to design every semi-automatic pistol to be made by the Colt factories and variations of his principles have appeared in millions of other guns produced in factories all around the world.

Having proved that semi-automatic operation was a workable principle, designers turned their attentions to evolving better applications. Thus it was that the majority of nations involved in World War 1 had adopted a semi-automatic pistol. For instance, the Americans, before their entry into the war, went through a great deal of heart searching following experiences in combat against the Moros in the Phillipines, where troops had been badly let down by the 0.38in service revolver. They conducted an exhaustive series of comparative tests to select a more suit-able hand gun. The ultimate winner was the Model 1911 Colt which, after minor alterations had been made to the basic Browning design to simplify production, was to become one of the best, most successful and durable of all combat handguns.

The Germans, of course, had equipped their troops with self-loading pistols well before the start of hostilities, the standard handgun being the 'Parabellum Pistole '08', with considerable numbers of 'Broomhandle Mausers' as back up. Although there were better semi-automatic weapons than the 'P '08', the German Army initially retained it as the basic service pistol despite the fact that it had been shown to be unsuited to combat conditions. In addition, like the Mauser, it was complex and expensive to manufacture as well as being too prone to variations in ammunition.

World War 2 saw the pistol as an item of military equipment eclipsed by other, newer types of weapon for there were greater and more important calls upon the precision manufacturing capacity required to build the handguns and there were no wartime developments of pistols that saw major service and therefore all the handguns in use throughout the conflict had been developed and accepted before war was declared and remained virtually unchanged throughout.

The British, despite a general trend in other major armies towards the self-loading pistol, remained firmly committed to the revolver as its service handgun. The standard weapon, since 1887, had been the Webley revolver in various Marks but of very similar design. The typical Webley revolver design was a double action lock and a six chambered cylinder, housed in a frame that was hinged to move downwards. The forward part of the frame and the cylinder was held in place by a 'stirrup lock" a solid bar of steel which passed across the end of the top strap and which was released by a thumb operated lever on the left side. This arrangement is tremendously strong, even under heavy loads and this factor was important to the success of the revolver at a time when huge calibres and massive 'stopping power' were in vogue. In fact the Webley revolver was one of the most reliable and most powerful service revolvers ever built. Operating the release lever and breaking the action ejected spent cartridge cases automatically.

Below:

For Britain and the Commonwealth, the revolver reigned supreme. Limited issue was made of the 0.455in Webley & Scott MkVI.

An interesting diversion from the main stream of this tale (for it was never an official issue weapon) was the Webley-Fosbery self-cocking revolver. Purchased in quite large numbers by officers during World War 1, the Webley-Fosbery was a semi-automatic revolver. Basically a hinged frame Webley revolver, the cylinder and barrel were mounted as a single unit which was free to slide along the top of the butt unit. Zig-zag grooves were cut in the cylinder and these engaged a stud fixed in the frame. Once loaded, the hammer was cocked manually for the first shot, the recoil from which drove the cylinder back along the frame and then forward again under return spring pressure. During this movement the hammer was cocked and the zig-zag grooves rotated the cylinder to present a new round for firing. Although an effective weapon, the mechanism of the Webley - Fosbery was susceptible to dirt and therefore proved unsuitable for trench warfare.

● **.455in Webley Revolver No1 MkVI** This, probably the best known of all British service revolvers was introduced in 1915 and was the last to be issued in the Victorian .455in calibre, developed specifically for use in the native wars of the late 19th century, before the niceties of the Geneva Convention ruled that only jacketed bullets could be used in 'civilised' warfare and where only a massive slug of soft lead could be relied upon to stop a frenzied enemy. In fact the 'MkVI' was not greatly different from the previ-

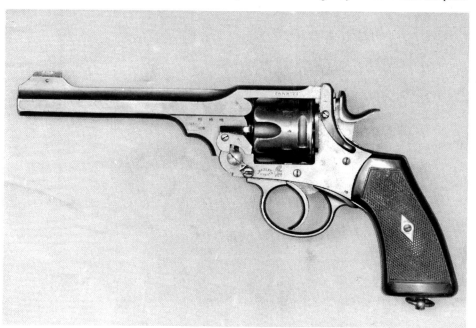

ous Marks with only the butt changing from the previous 'birds head' design to a more conventional squared look. The 'MkVI' saw service right through World War 1 and, although officially replaced in 1932, many were actually still in use up to 1947 with rear echelon troops. During World War 1, a short bayonet was developed to equip the pistol for close quarter fighting. Although this, together with a detachable shoulder stock, were never official issue, many were bought privately by officers. Some 'No1 MkVI' revolvers were rechambered for the .45in Colt Automatic Pistol cartridge and these rimless cartridges were held in the chamber by a pair of 'half-moon' spring clips holding three rounds each.

● **0.38in Webley Revolver Mk4** The .455in Webley, while an excellent pistol was, because of its violent recoil, only really effective in the hands of a well trained marksman. Because of this, the army researched pistol ballistics and arrived at the conclusion that a 0.38in, 200 grain bullet would be sufficiently lethal for its purposes and, because of its lighter recoil, be more likely to hit its mark in the hands of hastily trained troops. Webley & Scott had been developing a revolver for police use in this calibre and it was submitted for trial. The army however decided not to take up this weapon and to develop its own design and so the Webley was rejected for service use although it did continue as a commercial enterprise. By 1941, with an expanded British army, the Royal Small Arms Factory at Enfield was unable to keep up with demand and the Webley was introduced into service. Although up to Webley's usual high standards of workmanship, this model was of wartime finish. This pistol, like the 'No1 MkVI' was manufactured by Webley & Scott Ltd., BirminghamM

● **0.38in Enfield Revolver No2 Mk1** The exterior design of this weapon was very much in the Webley & Scott pattern but the lockwork was unaccountably changed to make the trigger

Below:
When a lighter calibre was needed and the 0.38in Webley MkIV was issued. Illustrated is a short-barrelled version.

Top right:
Expansion of forces necessitated the development of a pistol with simpler lockwork — the No2 Mk1 was built an Enfield.

Bottom right:
For some troops, the hammer spur was found to catch in clothing and the No2 Mk1* was developed with the spur removed and capable of only double-action operation.

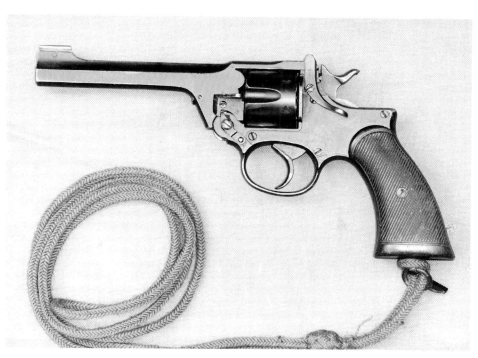

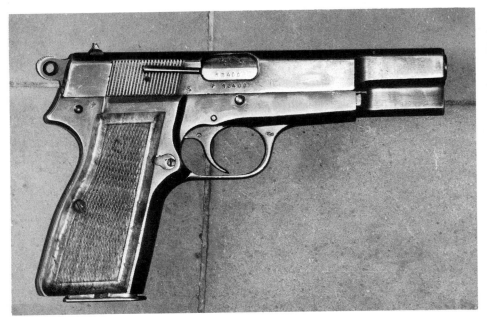

action very stiff and therefore difficult to fire accurately. As 'Mk1' pistols were withdrawn for overhaul or repair they were modified to 'No2 Mk1*' pattern and unmodified examples are comparatively rare.

● **0.38in Enfield Revolver No2 Mk1*** Like its Webley predecessors, the Enfield was capable of either single or double-action operation. However some categories of troops, particularly those, such as tank crews, working in enclosed conditions, found the exposed hammer spur prone to snagging. In 1938 the 'Mk1*' was introduced, with the hammer spur removed. This prevented thumb cocking of the hammer and limited the weapon to double-action use only. The trigger pull was lightened by two pounds and the grip plates were altered to provide better grip. The right grip was provided with a brass disc upon which were stamped regimental identification numbers.

● **0.38in Enfield Revolver No2 Mk1**** At the end of July 1942 a final modification was made to speed manufacture. The hammer stop of the 'Mk1*' was left out and minor modifications made to the lockwork. Leaving out the hammer stop made the pistol prone to firing if dropped and, while this was acceptable in wartime, the modifications were reversed and most models were converted back to 'Mk1*' pattern once the peace was signed. In addition to the Royal Small Arms Factory at Enfield, these pistols were made by Singer Sewing Machine Co at Clydebank and Albion Motor Co at Glasgow, Scotland.

Above:
Eventually, progress was forced upon the British Army and the Browning-designed 'HP' semi-automatic pistol was introduced.

● **0.38in Smith & Wesson** Losses of weapons at Dunkirk and expanding requirements forced the British Government to purchase pistols from abroad and, to help meet this need, Smith & Wesson built a revolver chambered for the 0.38in British Army cartridge. Known as the model '38/200" or 'British Military' it was a version of the company's commercially produced 'Military and Police' model. Early examples were of the normal commercial finish, with chequered walnut grips which bore the Smith & Wesson medallion and were supplied in four, five or six inch barrels. Later models were sandblasted with plain walnut grips and produced in only a six inch barrel length. Unlike the Webleys and Enfields, this pistol was light and accurate and was popular with the troops. Its one drawback was a weak mainspring which, with use, sometimes failed to provide enough power to the hammer to set off the cartridge primer.

● **9mm Browning (FN) HP, No2 Mk1*** This was the only semi-automatic pistol to be regular issue in the British Army during the war but only to airborne and commando units. The pistol had been designed by John Moses Browning in the 1920s and developed by Fabrique Nationale in

Belgium. It was adopted by the Belgian Army as the 'Browning High Power' or 'HP 35' in 1935. When war was declared, the Belgians sent all the drawings to England but, as there was no need or manufacturing capacity for the weapon, the drawings were sent on to the John Inglis company in Canada and the pistol went into production there.

The Belgian FN factory continued to manufacture the pistol during occupation and most examples were issued to SS and paratroop units as the 'Pistole Modell 35(b)'. Many of these occupation models are believed to have been sabotaged and are therefore treated with caution by shooters. The original design had a tangent rear sight graduated to 500yd and came with a wooden holster which doubled as a clip-on stock. Inglis supplied weapons of this specification to the Chinese Nationalist forces but those issued to Canadian and British troops had a simple fixed rear sight. The Browning was a very popular weapon, not least for its unusual 13-round magazine capacity. The wartime models in British service were manufactured only by John Inglis & Co of Toronto, Canada.

GERMANY

The Germans had been pioneers of the self-loading pistol and therefore it is not at all surprising that they should have three very good, well proven and distinctly different examples of the species in their armouries at the outbreak of war; weapons that were to become probably the most prized souvenirs for returning servicemen. This statement simplifies the supply situation greatly

for, between 1914 and 1945, the Germans had almost 30 different models of home designed and built handguns approved for service and, during World War 2, adopted a great variety of pistols captured from enemies or built in countries under occupation. Despite this profusion of types, the official issue handgun of the German army throughout World War 2 was the Walther 'P38', supplemented by large quantities of the Parabellum 'P '08' or Luger and smaller quantities of the Mauser 'Model 1934.' Of these, the Walther 'P38' was, despite the mythology and mystique that surrounds the Luger, undoubtedly the best and remains in volume production to this day. The Germans in fact had an embarrassment of riches and other well known and respected weapons such as the Walther 'PP' and 'PPK'; the Mauser 'HSc' and the Sauer 'Model 38' also saw extensive wartime service although were never infantry weapons.

● **Pistole Parabellum Modell 1908 (P '08) 'Luger'**
The 'P '08' was evolved from the earlier Borchardt design by Georg Luger at the turn of the century. Luger refined the cumbersome and complex mechanism of the Borchardt. He retained the toggle link locking mechanism of the Borchardt but replaced that gun's large, clock-like mainspring with a leaf spring in the grip which altered the weapon's profile and in the process

Below:
The Germans were firmly entrenched with the self-loading pistol. Early issue was the P '08 'Luger'. Standard issue was the four inch barrelled version.

Above:
The 'Luger' was soon replaced by the easier-to-manufacture and better Walther 'P38'.

made it a much better balanced, handier weapon. Original weapons were made in 7.65mm calibre but, to make the gun more suitable for military use, the calibre was changed to 9mm and the leaf mainspring replaced by a compact coil spring.

Despite the mystique that surrounds this pistol — it does have an attractive appearance, good balance and 'pointability' — it was not a particularly good military weapon for a number of reasons. The 'P '08' was very expensive to produce with large numbers of small parts, each requiring careful manufacture and fitting; the strength of the springs was critical and again had to be very carefully produced and the toggle was too dependent on standard quality ammunition with under or over powered rounds causing malfunctions. it also suffered from the same failing as the Webley-Fosbery, being prone to debris into the mechanism. Although there were better weapons of the type around than the 'P '08', the German Army adopted the pistol and retained it as the basic service pistol until 1942 when manufacture finally ceased with a total of probably three million having been made. The normal issue model was 9mm calibre with a barrel length of four inches and a magazine capacity of eight rounds. It was issued throughout the German Army as an officer's sidearm and also to machine gun and artillery crews, despatch riders, signallers and non commissioned officers but without the variations of barrel length that had occurred during the World War 1. Like many other German weapons, those produced early in the war were

of excellent quality, the standard of which decreased as the war progressed. The 'P '08' was manufactured during the war by DWM, Berlin; the Royal Arsenal, Erfurt; Simson & Cie and Heinrich Krieghoff Waffenwerke, Suhl; and Mauserwerke AG, Oberndorf.

● **7.63mm Mauser Modell 1912** In the early 1890s Paul Mauser, who had until then paid little attention to the pistol market, saw the possibility of adding lucrative military pistol contracts to his domination of the rifle field and adopted a self-loading pistol developed by two of his employees. The result was the Mauser 'Model 1896' a large, reliable, accurate and beautifully made pistol chambered for a powerful bottle-necked 7.63mm cartridge. Not a single pin or screw was used anywhere in the mechanism, the whole assembly fitting together like a jigsaw puzzle. The pistol was equipped with a wooden holster which clipped on to the butt to double as a shoulder stock. This was useful for the 7.63mm round produced a very high muzzle velocity and a flat trajectory that made it possible for a good shot to use it accurately at up to 1,000yd. The pistol was widely used in World War 1 and the design was resurrected in 1932 for certain sections of the German armed forces although its

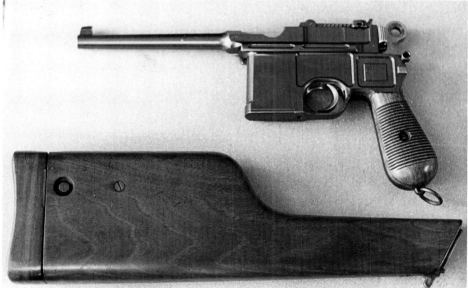

Top:
Some units were issued with the venerable Mauser 1912, the 'Broomhandle'.

Above:
The wooden holster, clipped on to the butt, made the pistol a fairly effective light carbine and helped steady the selective-fire model (1932 version) — the 'Schnellfeuer'.

construction did not lend itself to mass production. During World War 2, the Mauser was most widely used by SS and Feldgendarmerie units. Some examples were converted to fire 9mm Parabellum ammunition and these are recognised by a large '9' carved into the grips and filled with red paint.

● **7.63mm Mauser Modell 1932 (Schnellfeuerpistole)** The 1912 model pistol had been widely copied in China and Spain between the wars and one variant that appeared in Spain incorporated a selector mechanism allowing the weapon to be fired as a fully automatic. To increase the ammunition capacity, a detachable 20-round magazine box magazine replaced the integral 10-round magazine of the original. Prompted by this plagiarism, Mauser produced its own version of the copy but much better engineered — the 'Modell 1932'. Although the 'Modell 1932' was a poor substitute for a proper submachine gun, as its light weight and high rate of fire made it difficult to control, it was used extensively on the Eastern Front.

Because the design was derived from a semi-automatic pistol, it fired from the closed breech so that when the the trigger was released at the end of a burst, the bolt closed on a fresh round in a hot chamber. As few as 30 shots would produce enough heat in the chamber to 'cook-off' and fire the cartridge spontaneously within seconds. All the Mauser pistols, which throughout production managed to retain their high standards of finish, were produced by Mauserwerke AG at Oberndorf.

● **9mm Walther Pistole Modell 1938** The firm of Carl Walther had been producing high quality pocket pistols for many years before deciding to enter the military pistol market in the early 1930s with a design for a weapon in 9mm Parabellum. In a complete change of direction from their previous weapons they produced a recoil operated pistol with barrel and slide locked together by a wedge until chamber pressure dropped to a safe level, when the slide was allowed to continue rearwards to eject the spent case, recock the mechanism and chamber a new round. The pistol had an internal hammer and a double action trigger mechanism. This mechanism allowed the pistol to be safely carried with a round in the chamber and the hammer down. To fire, all that was required was for the firer to pull the trigger which raises the hammer and releases it at full cock to fire the cartridge. From then on the hammer remained cocked and the pistol functioned as a normal self-loading pistol. After military testing a few minor changes were called for, the main one being the request for an external hammer for safety reasons and in 1938 it was adopted by the Army as 'Pistole '38' or 'P38'.

A key advantage of the weapon was that it permitted a degree of mass production in its manufacture unavailable with other designs. At peak production, three factories were producing 'P38s' with countless sub-contractors producing parts for assembly. The pistol performed well in even the most adverse conditions and most wartime models were sound, the main exceptions being those of 1944 onwards which were made from poor materials which wore out quickly and became unsafe. The 'P38' was manufactured by Carl Walther, Zella-Mehlis; Spreewerke, GmbH, Berlin; and Mauserwerke AG at Oberndorf.

Above:
The Italian 'M1910' 'Glisenti' was the country's most powerful pistol.

The Italian Army was only slightly better served by handguns than by other small arms. This is not really saying very much because, with the notable exception of their submachine guns, Italian small arms of World War 2 were a fairly abysmal collection showing little imagination, appreciation of developments elsewhere in the world or consideration for the soldiers who had to use the equipment provided.

● **9mm Pistola automatica M1910 (Glisenti)** Adopted just before the outbreak of war, the actual origins of the 'Glisenti' pistol are a little vague. Although often attributed to the designer Revelli, its real origins are probably in the Swiss Haensler & Roch pistol of 1905. Unfortunately, the 'Glisenti' is structurally and mechanically weak and is chambered for a special 9mm cartridge which, while being an exact dimensional duplicate of the 9mm Parabellum, is loaded to a much weaker level. This is a potentially dangerous situation as inadvertent use of 9mm ammunition would exceed the very slim safety margins of the weapon with disastrous results and this is probably the reason that few 'Glisenti' pistols seem to have survived the war.

● **9mm Beretta, Model 1934** This was undoubtedly the most successful of the Italian service pistols and was derived from a design, produced by Pietro Beretta in 1915, for a simple and reliable blowback pistol. Unlocked blowback operation was made possible by the choice of cartridge for, although the calibre was 9mm, it was not the ubiquitous Parabellum but a shortened, weaker round, also known as the '0.380in Automatic'. The 'Model 1934' was simple in design but also extremely strong as well as being very compact and these virtues made it popular, not only with the Italian Army but also as a souvenir among Allied troops. This was despite the fact that the ammunition lacked power or accuracy, a situation that was exacerbated by a very short barrel which soon went out of alignment in use. The compact design of the pistol also made for a small grip and to improve this feature, many magazines were fitted with a curved lower extension. A slight oddity of the design was that the action was held open by the magazine platform when the last shot was fired (a feature which sometimes made magazine changing difficult) but slammed forward on the empty chamber when the magazine was removed, making manual recocking necessary. Military issue examples were usually marked with the letters 'RE' under a crown. The manufacturer was Pietro Beretta, Gardone.

JAPAN

Where in most other armies of the World War 2 the place of the pistol was largely taken over by the submachine gun, this was not the case with Japan. The most likely reason is that they never managed to develop a submachine gun that proved a successful alternative but added to this must be a large element of tradition that demanded that officers carry swords and pistols.

In fact, beginning in 1897, Japan was among the first countries to institute a programme to

Above:
The 'Glisenti' was largely replaced by the Beretta 'M1934', firing the relatively feeble 9mm short round.

develop a military semi-automatic pistol, its designer, Col Kirijo Nambu demonstrating the weapon to the Emperor at the Toyama Military Academy in 1909.

● **8mm Taisho '04 (Nambu)** This pistol is often referred to as 'the Japanese Luger' but, although there is a superficial resemblance, the two weapons are not at all alike mechanically and the 'Nambu' is an original design, although it does contain elements of similarity to the Italian 'Glisenti'. The designation of the weapon is taken from the fourth year of the Taisho era (1915 by our calendar). The 'Nambu' has a barrel and extension moving on top of the frame and is cocked by pulling back a pair of 'ears' on the rear of the bolt. Oddities of the design are that it can be reassembled without the breech block in place, a situation making it extremely dangerous to fire and the use of a single, small recoil spring in a recess on one side of the receiver, which gives it a 'lopsided' appearance. There was no manual safety catch on the pistol although a grip safety was let into the front of the grip, just below the trigger guard. The pistol fired a special but underpowered bottle-necked 8mm cartridge. While it was never an officially issued weapon, it was privately purchased by officers and very widely used throughout the war. A scaled down version in 7mm calibre and popularly known as the 'Baby Nambu' was also produced and, was popular with senior officers.

● **8mm Taisho 14** In 1924, the Arsenal in Tokyo which had produced the original 'Nambu' pistol was destroyed by an earthquake and the Japanese

took the opportunity to develop a modified design for manufacture at the Nagoya Arsenal. Most of the work was handled by the Government design office with advice from Col Nambu and the intention was to produce a pistol that would be easier to manufacture. The exterior appearance remained very similar but with a manual safety catch above the trigger guard in a position which required the shooter to operate it with his free hand. Two recoil springs were used and gave this version a more balanced appearance than its predecessor. The 'Taisho 14' was adopted as service issue for NCOs in 1925 or the 14th year of the Taisho reign. A 'Baby Nambu' version, again in 7mm calibre, was adopted by the Air Force. In the early 1930s a version was developed for use in Manchuria. This has an enlarged trigger guard to permit its use with gloved hands and is unofficially known as the 'Kiska Model' from its first discovery by Allied troops during the campaign in the Aleutian Islands.

● **8mm Type 94** The previous 'Nambu' pistols were, despite their lack of power, workable and reasonably efficient pistols, although rather heavy. Officers, who were required to buy their

Below:
Never officially adopted by the Japanese military, the Taisho '04 pistol was nevertheless carried by thousands of officers.

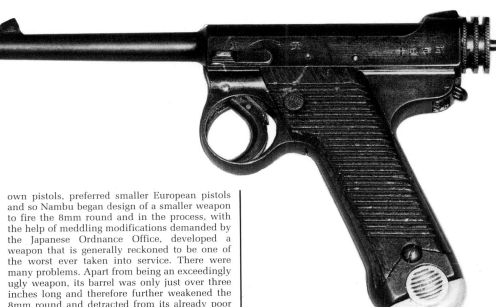

own pistols, preferred smaller European pistols and so Nambu began design of a smaller weapon to fire the 8mm round and in the process, with the help of meddling modifications demanded by the Japanese Ordnance Office, developed a weapon that is generally reckoned to be one of the worst ever taken into service. There were many problems. Apart from being an exceedingly ugly weapon, its barrel was only just over three inches long and therefore further weakened the 8mm round and detracted from its already poor accuracy. Mechanically, it was possible to release the firing pin and fire the weapon before the barrel and breech were locked together. Perhaps worst of all, the sear had an exposed metal extension strip on the side of the frame and so, simply

Above:
The Taisho '14 was an attempt to improve and simplify the previous design.

Above:
Although similar in outline, there were basic differences between the two models. The Taisho '04 has a grip safety catch, an adjustable rear sight, distinct 'ears' on the bolt and a shaped butt.

Above:
On the simplified Taisho '14, the safety catch has been removed and the butt straightened. The rear of the bolt is now a series of milled rings. The standard of finish was usually rougher than on the older model.

Above:
Quite possibly the worst military pistol ever issued, the Japanese 'Type 94' was positively dangerous to the user and few of the model reach the surplus or deactivated market.

depressing this by handling the pistol carelessly, would cause it to fire. Strangely, the 'Type 94' seems to have been popular with its users. All Japanese pistols were manufactured at State Arsenals.

SOVIET UNION

Once the Soviet Union had recovered from the attrition of the World War 1 and the effects of the subsequent Revolution, it began to rebuild the Red Army and began to look for new weapons. Among these, a more modern replacement was required for the ancient and cumbersome Nagant revolver and it was decided to build any new pistol around the 7.63mm Mauser automatic pistol cartridge, which had become familiar to the Russians through the large numbers of such pistols imported into the

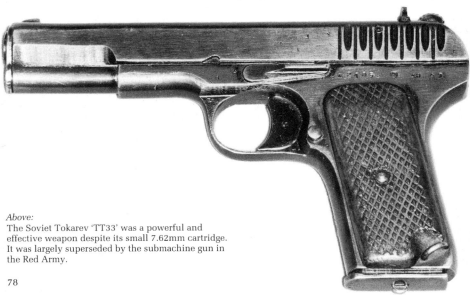

Above:
The Soviet Tokarev 'TT33' was a powerful and effective weapon despite its small 7.62mm cartridge. It was largely superseded by the submachine gun in the Red Army.

country in the 1920s. The reason for this decision was that the Artillery Commission, whose responsibility such a decision was, believed it would make for a powerful pistol round as well as being suitable for use in submachine guns. In fact, the submachine gun would eventually virtually replace the pistol in Soviet service apart from limited issue to vehicle crews and staff officers.

● **7.62mm Tokarev TT33** In 1930, Feodor Tokarev presented a design for a semi-automatic pistol, closely based upon Colt-Browning designs, using a swinging link system to engage lugs on the barrel with matching grooves cut into the slide to lock breech and barrel together. Designated the 'TT33' — the initials from Tula, the arsenal at which the pistol was built and Tokarev, the designer — the pistol had certain improvements over the Colt-Browning such as having machined lips — normally part of the magazine — in the frame to improve reliability. The hammer and mainspring could also be easily removed, as a unit, for cleaning. A notable feature of the pistol is the complete absence of any form of safety catch. There were in fact two versions, externally identical but in the 'TT33' locking is achieved in the same way as for the Browning design with two lugs on the top of the barrel only, while in the 'TT34' the lugs run completely around the barrel to allow them to be manufactured in a simple lathe operation rather than having to be milled out and thereby simplify production. Using the Russian 7.62mm variant of the Mauser round the Tokarev was a powerful pistol which, coupled with its lightness, made recoil rather violent.

The pistol, particularly the revolver, is very much part of the American heritage and features prominently in its history. Therefore, during the early part of the 20th century, while development of the semi-automatic pistol had been progressing apace in Europe, similar development on the other side of the Atlantic was very slow with John Moses Browning alone among well known designers to show an interest. In 1895 he took out a number of patents on blowback pistols and tried, without success, to interest the Colt company in these designs. Their only professed interest was in a heavy calibre model that would be suitable for military use.

He eventually went to Fabrique Nationale in Belgium to perfect his designs and to produce several commercially successful blowback pistols that provided the funds to allow him to work on a locked breech design that would fulfil Colt's requirements. The system he perfected is often called the 'swinging link' and in his first successful design the barrel was supported in the frame by two hinged links, one attached to the muzzle and one to the breech. The barrel was surrounded by a slide, in the underside of the top surface of which was cut a pair of slots which matched two lugs machined on top of the barrel. As the slide moved forward it collected a round from a magazine in the butt and fed it into the chamber. As the breech closed, the barrel was moved forward and, because of the links, as it moved forward it also rose and the lugs located into the slots on the slide, locking the action. The action was reversed as the assembly moved back-

Above:
Probably the best military pistol of the war was the Browning-designed Colt 0.45in 'Model 1911A1' which was only relatively recently replaced as service issue.

ward in recoil. The principle was so successful that almost half the military pistols in service in the world use it or a variant of it.

● **.45in Automatic Pistol M1911A1** In 1908, the American Army conducted a series of trials to arrive at a new service pistol. A minimum calibre of .45in was stipulated and, among weapons in competition, were the Webley-Fosbery and the Parabellum ('Luger') uprated to a .45in calibre. The tests were exhaustive and, at the end, a Colt weapon, designed by John Browning was the winner and taken into service, with slight modifications to meet the Army's demands for safety and simplicity, as the 'M1911'. The main change was to use one link beneath the chamber, rather than the two in the original Browning design. The second link was replaced by a bearing bush at the muzzle. In addition, a manual safety catch was fitted in the frame and a grip safety in the rear edge of the butt.

Following combat experience in World War 1, some detail changes were made, the shape of the butt was altered to provide a better fit for the hand, the trigger was shortened and the front edge of the butt was champfered away behind the trigger to allow the trigger finger a better grip and the hammer spur was shortened. Firing a 230 grain bullet at 860ft/sec, the 'M1911A1' was, for much of its service life, the most powerful service pistol in the world although the reverse side of this is that it was difficult to shoot accurately without extensive training. Despite this the 'M1911A1' remained in service until 1985. In addition to manufacture by Colt Patent Firearms Manufacturing Co, the pistol was made by Remington Arms, Bridgeport, Connecticut; Springfield Armoury, Springfield Massachussetts; Union Switch & Signal Co, Swissvale, Pennsylvania; Ithaca Gun Co, Ithaca, New York; and by Remington Rand Inc, Syracuse, New York.

Pistol Technical Data

Britain & Commonwealth

Type	Calibre	Weight	Action	Overall length	Barrel length	Feed
REVOLVER No1 MkVI	0.455in	2.37lb	Double action	11.25in	6in	6 chamber cylinder
REVOLVER Mk4	0.38in	1.68lb	Double action	10in	5in	6 chamber cylinder
REVOLVER No2 Mk1	038in	1.81lb	Double action	10in	5in	6 chamber cylinder
REVOLVER No2	0.38in	1.68lb	Double action	10in	5in	6 chamber cylinder
REVOLVER No2 Mk1**	0.38in	1.68 lb	Double action	10in	5in	6 chamber cylinder
0.38in SMITH & WESSON	0.38in	1.5lb (6")	Double action	10.125" (6")	4in. 5in or 6in	6 chamber cylinder
9MM Browning (FN) HP No2 Mk1*	9mm	2.18lb	Semi-automatic	7.75in	4.65	13-round box magazine

Germany

Type	Calibre	Weight	Action	Overall length	Barrel length	Feed
9mm P'08	9mm	1.87lb	Semi-automatic	8.75in	4in	8-round box magazine
7.63mm MAUSER1912	7.63mm	2.75lb	Semi-automatic	11.75in	5.5in	10-round integral
7.63mm MAUSER 1932	7.63mm	2.90lb	selective	11.75in	5.5in	0 or 20-round box
9mm P38	9mm	.84lb	Semi-automatic/ double action	8.38in	5in	8-round box

Italy

Type	Calibre	Weight	Action	Length	Barrel length	Feed
9mm M1910 (GLISENTI)	9mm	1.90lb	Semi-automatic	8.25in	3.5in	7-round box
9mm BERETTA M1934	9mm	1.46lb	Semi-automatic	6in	3.75in	7-round box

Japan

Type	Calibre	Weight	Action	Length	Barrel length	Feed
8mm TAISHO '04	8mm	1.96lb	semi-automatic	9in	4.7in	8-round box
8mm TAISHO 14	8mm	1.97lb	semi-automatic	8.95in	4.75in	8-round box
8mm TYPE 94	8mm	1.75lb	semi automatic	7.125in	3.125in	6-round box

Soviet Union

Type	Calibre	Weight	Action	Length	Barrel length	Feed
7.62mm TOKAREV TT33	7.62mm	1.81lb	semi-automatic	7.68in	4.57in	8-round box

United States

Type	Calibre	Weight	Action	Length	Barrel length	Feed
0.45in M1911A1	0.45in	2.46lb	semi-automatic	8.5in	5in	7-round box

5

Machine Guns

Above:
In the northern European theatre, a British soldier uses rubble as cover from which to use his 'Bren' gun. *IWM/B5382*

At the onset of World War 2, the automatic machine gun was barely 50 years old and yet, in that relatively brief history, it had changed the face of war and the way in which it was waged; made traditional tactics obsolete and created new ones in their place.

The father of the modern machine gun was Hiram Maxim, born in America in 1840 but of French parentage. First apprenticed as a coach builder and then having worked in engineering works and shipyards, Hiram Maxim interested himself in the new science of electricity. Legend has it that while visiting an electrical exhibition in Paris in 1881, he was advised by a friend 'Hang your electricity, if you wish to make your fortune, invent something that will allow the fool Europeans to kill each other quicker!' He took the advice to heart; went to London and spent two years analysing the methods of operation of contemporary firearms. The machine guns of the day were mechanically operated, like the Gatling gun, which used multiple barrels, rotated by a hand crank or a bulky electrical system, firing cartridges fed from a hopper.

Maxim soon discovered that large amounts of energy were released by the firing of the cartridge, most of which was being wasted producing recoil, gas or heat and only a small proportion used to propel the bullet. He deduced that of these areas of waste, the force of recoil could be most easily harnessed to operate a machine gun mechanism. His first design was for the .45in Martini Henry cartridge and marked a complete departure from then current machine gun thinking. The mechanism was complex but did succeed in harnessing the recoil of the cartridge to power the mechanism; it also incorporated feed from a canvas belt of cartridges, entering the mechanism from the right of the receiver. The 'Maxim Automatic Gun' was demonstrated to an astonished press and military, one of whom commented that the gun would have a better chance of acceptance if it were made to allow essential parts to be stripped and replaced without special tools or training.

In 1885, Maxim demonstrated a new design that was so good in its basic principle that it remained in service with some armies, virtually unchanged, through to World War 2. At the heart of the simplified system was a 'toggle joint' — two solid metal members linked by a one-way hinge joint. The barrel of the new gun was extended into the body to form two side plates between which the breech block slid. When a round was chambered, the toggle joint formed a straight line linking the barrel and breech block.

Below:
A German machine gun team, one of whom has been given the unenviable task of acting as 'tripod'.
IMW/12789

When the cartridge was fired, the recoil force was transferred through the joint causing the barrel and breech block to move backwards together. After about half an inch of travel, a crank and roller applied force to the joint, breaking the straight-line and causing it to fold downwards about the hinge, allowing the breech block to continue travelling backwards while the barrel stopped. Eventually a spring halted and reversed the motion of the breech block, chambering a new round and allowing the toggle to lift once again to form a solid locking strut. During this action, a series of pawls lifted the the ammunition belt and presented a new round ready for loading.

Having proved the possibility of harnessing the power of the cartridge to work the mechanism of a gun, Maxim gained patents on all possible applications of the recoil principle of operation and competitors were forced to develop other systems. In America, John Moses Browning produced a design in which gas was vented from the muzzle into a cylinder below the barrel. Here, a piston was driven to the rear and operated a rod which unlocked the bolt to allow the bolt to be moved to the rear, extracting the used case and, when it was returned under spring pressure, picking up a fresh round. Gas operation, as this method is called, is really only suitable for short burst as otherwise fouling from burnt powder can build up to clog the port and cause the weapon to malfunction. For this reason, the use of gas operation is usually restricted to light machine guns.

The third major method of operation for machine guns is retarded blowback. This method of operation, which employs no method of locking the mechanism and which leads to a violent action, is best typified by the Austrian Schwarzlose, a heavy weapon that relied upon an enormous breech block and a strong spring, backed by toggle link which performed no locking function but was used to slow down recoil. Unlike the Maxim toggle, that of the Schwarzlose was not straight when the breech was closed but was folded back on itself. As the block moved it straightened but, because of mechanical advantage built into the lever system of the toggle, the straightening action caused a slight delay. The Schwarzlose, although a large and heavy weapon; because it employed the blowback principle, was also a very simple design. However, there is an inherent problem with retarded blowback. The gas pressure that pushes back the action and ejects the bullet from the barrel also pushes out the walls of the cartridge case in all directions, causing it to stick in the chamber, particularly at the front end where the case metal tends to be thinner. The result is often that the case splits, jamming the mechanism. To overcome this problem, the Schwarzlose had to incorporate an oil reservoir and pump in the mechanism to spray each round with a thin film of oil before it entered the chamber.

For sustained use, the recoil system has proven to be best, a fact that was proved dramatically by the British 100th Machine Gun Company on the Somme in 1916. The Company was tasked to deny the Germans a particular piece of the battlefield for 12 hours. During the crucial period, 10 Vickers guns fired just short of one million rounds at a range of 2,000yd. To keep the guns cooled all the normal cans of water, the waterbottles of the entire Company plus the contents of all the urine buckets in the surrounding area were emptied. The guns wore out almost 100 barrels and were evidently firing as well at the end of the encounter as at the start.

This action was perhaps typical of the use of the automatic machine gun in the early part of its history where its capabilities were employed more in the role of an artillery piece than as an infantry weapon. The early water-cooled guns, such as the Vickers, Schwarzlose and Maxim — heavy bulky and lacking manoeuvrability — were particularly suited to this type of work and were carefully dug into semi-permanent positions, providing interlocking fields of fire to defend areas of territory. In World War 1, used from this type of position against slow moving infantry, advancing across rough ground, their effect was devastating. As a result the machine gun gained a terrible reputation and, with artillery, was largely responsible for the static nature of the war.

Contrary to legend, World War 1 did not begin with the armies massively equipped with machine guns. None of the early participants were particularly enthusiastic about the weapon in the early stages and, in 1914, Britain, France

Left:
Machine gun development and usage came on in leaps and bounds. The static conditions of World War 1 had seen them used almost like artillery, a fact illustrated by this complex sight from a Vickers gun.

Below:
Construction also changed, from the heavy precision machining used in the Vickers ...

Bottom:
... through the lighter, but still precise, machining of the 'MG34' ...

and Germany had an almost identical ration of machine guns to troops — something like two guns per infantry battalion. Only when the trench warfare began did the Germans move machine guns from reserve battalions to the front line and begin the escalation. From that point all nations began to produce machine guns as fast as their industrial resources would allow which was not particularly fast as the current guns were complex masterpieces of specialist engineering, requiring many hours of skilled work. Under such pressure, the search therefore began for machine guns which could be produced more quickly and more cheaply.

Below:
... to the pressed steel construction of the German
'MG42' 'Spandau'.

Right:
Warfare was more fluid and demanded mobility.
Heavy tripods were replaced by the folding bipod.

Far right:
Where belts were retained, they were often coiled
inside metal boxes clipped to the side of the gun.

At the same time the realisation dawned that the heavy water cooled guns were not the entire answer to battlefield requirements. What was needed was a lighter model that could be relatively easily taken forward in attack; which could be put in place inconspicuously and got into action quickly. There was also pressure for a lighter machine gun for use in the aircraft of the emerging air forces. The concept of the light machine gun had been born.

The parameters set for the new type of weapon were not easy to meet. A prime requirement was for an efficient yet portable method of cooling, for the rapid fire of the machine gun soon raises the temperature of the barrel to a point where erosion becomes rapid. For instance, firing a relatively few belts of ammunition through a Vickers gun produced sufficient heat to boil the seven pints of water in the cooling jacket. The rate of fire was also a delicate balance between a high

Right:
Machine guns became lighter, many using quick-change box magazines rather than belts for ammunition feed.

rate that quickly produced overheating and a slow rate that made for a heavy and inefficient weapon. The final problem was a method of feeding cartridges to the weapon that would be light and efficient but not vulnerable to the dirt of the battlefield.

The almost universal solution was a gas operated weapon with a simple bipod mounting; that could be carried and operated by one man and on which the barrel could be rapidly and easily replaced. Cartridge feed was from a magazine holding 30 or more rounds of ammunition. Perhaps typical was the Lewis gun which could be built for a small fraction of the cost of a Vickers in time and money. The Lewis carried its ammunition in a circular magazine on the top of the gun. Cleverly, the hot gases from the exploding cartridge were used to cool the gun. As the gases exited the muzzle they expanded in a fan shape, striking the front of an aluminium casing around the barrel and creating a vacuum which caused a continuous stream of cool air to be drawn through the fins of the casing from the rear of the gun. At 27lb, the Lewis was comparatively light and it was compact and so did not make an easy target for enemy counterfire. On the minus side, the gun was prone to a bewildering array of potential stoppages that necessitated the Lewis gunner to be dexterous and cool under fire in order to keep it operational.

The Lewis was an exception among the small number of light machine guns that were produced during the World War 1 for most were not particularly good or effective but they did set the standard and provide the nucleus for continuing developments during the interwar years.

BRITAIN AND COMMONWEALTH

Britain had made few advances in automatic weaponry between the wars; pressure for change had diminished with the end of hostilities and largely because huge stocks of machine guns remained from World War 1 and so the mainstay of the army's machine gun equipment was still the Vickers 'Mk1', although a minor reorganisation had removed many of these from the general infantry battalions into designated machine gun battalions. For general infantry support, the Lewis gun was retained, until well into the 1930s, despite its shortcomings.

● **0.303in Vickers Mk1**
The British Army had been one of the first to adopt the Maxim and used it, unchanged for almost 20 years. In the early part of this century, the Vickers company bought out Maxim and acquired rights to the patents of the machine gun which they promptly began to modify slightly. Careful recalculation of the stresses on the gun's mechanism allowed surplus material to be trimmed away and, in some instances, better quality or lighter metals to be substituted which resulted in a loss of something like one third of the original weight of the original Maxim. Vickers also redesigned the toggle link so that it now broke upwards. These minor changes improved significantly the performance of the gun and it was accepted for army service in 1912.

The Vickers was mechanically very similar to the original Maxim guns; recoil operated with the recoil force augmented by a muzzle attachment to deflect some of the gas rearwards to thrust on the breech mechanism. As the lock moved to the rear it collected a round from the canvas belt and stretched a return (or Fuzee) spring housed in a box on the side of the gun. Eventually the spring moved the bolt forward again to allow the cycle to repeat. The barrel was cooled by water contained in a surrounding thin metal jacket and, as this water would heat and evaporate at a pint or so for every 1,000-rounds fired, the gun was fitted with a condenser tube, draining into a petrol can to allow it to be reused.

Above:
Many weapons used a muzzle recoil-intensifier such as this on the Vickers in order to improve reliability.

The Vickers Gun remained virtually unchanged throughout its service life, last seeing action in the Radfan in 1968. Examples may be found with either plain or fluted cooling jackets, although this does not indicate any difference in Mark of gun. That the Vickers remained is service so long is a tribute to its reliability, for even when it did stop firing, the fault could usually be cured quickly. This reliability is due in no small part to the way in which it was built with precisely machined and fitted components made from high quality materials. Vickers guns were manufactured by Vickers Son & Maxim at Crayford in Kent and by various Royal Ordnance Factories. Examples may also be found manufactured by Colt in the United States and supplied during the early part of the war to supplement domestic production. These were originally made for the US forces in World War 1 and chambered for their 0.30in-'06 service round. Often issued to Home Guard units, various parts were daubed with red paint to warn against trying to use 0.303in ammunition.

● **BREN Light Machine Gun** It is surprising how often, when dealing with the history of a British Army weapon of World War 2, a story of unpreparedness, panicked development or acquisition and compromise is found. An exception — just — is the 'Bren' light machine gun, without argument, one of the best and most reliable weapons of its type ever developed. In 1932, a series of trials were started to find a successor for the Lewis gun. Hot favourite was the Vickers-Berthier, a French design, the rights of which had been purchased in 1925 and which had been in service with the Indian Army. However, in the middle of the trials, a report was received from the Military Attache in Prague describing, in glowing terms, the performance of a Czechoslovakian gun, the 'ZB 26', designed at the Zbrojovka factory in Brno by Vaclac Holek. At once an example, plus 10,000 of its 7.92mm cartridges, was obtained at a cost of just over $130 and included in the British trials.

The weapon was gas operated and acquitted itself so well in the trials that it was adopted for service straight away, with the proviso that the Czechs were to prove that the gun could work with the rimmed 0.303in British service round. Within an extremely short period they returned a version chambered for 0.303in, sporting the 30-round curved magazine that was to become one of the gun's most obvious characteristics and with the sights recalibrated in yards rather than metres. There were also other detail changes, including a shortened barrel, more suited to the characteristics of the propellant of the 0.303in cartridge and with the gas port moved back accordingly.

The gun was quickly accepted and the Government decided it would be manufactured by the Royal Small Arms Factory at Enfield and christened the 'Bren' (**BR**no + **EN**field = BREN). The Czechs quickly provided a full set of drawings in Imperial rather than metric measure and production began almost immediately. The work was complex for the body of the gun had to be milled from solid metal and, according to contemporary sources, this process required 270 machining operations, employing more than 550 gauges accurate to 1/2,000th of an inch. However, the first Enfield-built gun was completed in 1937 and by the summer of 1940, over 30,000 'Brens' had been issued — most of which were subsequently lost at Dunkirk. Production actually reached its peak of 1,000 a week in 1943. 'Bren' guns were also built by the Lithgow Small Arms Factory in Australia and by Inglis in Canada who also made examples in 7.92mm for the Chinese Nationalist Army. By the end of the war, more than half of all 'Bren' guns made had been built in Canada.

The 'Bren' is one of those very rare military weapons that commands nothing but respect and affection from those that have used it. The method of gas operation chosen by the Czechs incorporated a variable gas port system which allowed the gunner to compensate quickly and easily for variation in rate of fire simply by twisting an adjustable plug to any one of four positions. This also allowed provision to be made for dirt in the mechanism, fouling or ammunition of varying quality, any of which factors might be responsible for creating stoppages in a lesser gun. If problems were encountered, all the gun team had to do was release the barrel, rotate the plug through 90° to the next biggest port, fix the barrel and recommence firing, all of which could be achieved in less than 10sec. This speed was made possible by the quick change barrel system of the 'Bren'. The entire barrel assembly, including chamber, gas port and front sight could be removed simply by lifting a latch on the side of the receiver.

It was quickly found that although the curved magazine was designed to house 30-rounds, 28 was the optimum number and these had to be loaded carefully to ensure the rims were correctly located.

The receiver of the 'Bren' is a solid, one-piece, machined forging which incorporates the gas cylinder and magazine housing. From the gas cylinder at the front of the receiver a folding bipod is pivoted. On top of the receiver, the magazine housing has a sliding pressed metal cover to seal it when there is no magazine in place. The

ejection slot on the underside is similarly protected against dirt and debris and is held in place by a sprung catch which, while it has to be closed manually, releases automatically as the bolt carrier moves forward on firing. Because of the vertically mounted magazine, the sights were to the left. The foresight was on the front of the barrel on a winged extension arm. The rear sight fitted on the side of the receiver and, on early models was a complex drum and arm arrangement which was later replaced by a simpler leaf sight.

All furniture on the gun, from the carrying handle through the pistol grip to the stock was of wood and the short butt contained the mainspring spring and recoil buffer assembly.Total weight was a little over 22lbs and the 'Bren' was superbly accurate at ranges up to 1,000yd. The change lever allowed single shots as well as fully automatic fire and 'Bren' gunners were trained to use the single shot mode wherever practical, in effect making the 'Bren' a large self loading rifle. Accuracy was aided in this mode as, when the change lever was switched to 'R' for repetition, the trigger pull was altered to provide an almost two-stage let off. The 'Bren' was a superbly accurate weapon even in fully automatic mode.

A number of tripods were available to increase the versatility of the 'Bren' and some guns were modified to allow them to take a 100-round drum magazine in place of the conventional box, this was particularly used when the 'Bren' was employed in an anti-aircraft role.

● **Bren Mk1** This was the early Enfield-built model with radial sight, a distinctively shaped butt with folding shoulder support and with a wooden handle hinged beneath the butt.

Below:
Lighter, more mobile and much more widely used was the Czech-designed 'Bren'. In this illustration a Canadian built Mk1m of 1943 is shown.

● **Bren Mk1(M)** This Mark was built only in Canada by the Inglis Co and differed from the 'Mk1' by simplification of the butt shape and removal of the shoulder support and butt-mounted handle. Other minor differences include bipod legs that did not telescope and and enlarged gas vent in the barrel.

● **Bren Mk2M** A model built in both Canada and the UK and carrying the simplified butt of 'Mk1(M)' but with a flip up leaf rear sight in place of the drum.

● **Bren Mk3 and Mk4** Lightened versions of the Bren 'Mk1' & 'Mk2' (by some 3lb) with a shortened barrel.

● **Vickers-Berthier** Originally designed by Frenchman Lt Andre Berthier in 1920, rights to the gun were bought by Vickers and the gun offered commercially to several countries. In 1933, at roughly the same time it was being tested by the British, the Indian Government adopted it as their standard light machine gun. In appearance it is similar to the 'Bren', sharing the curved magazine placed above the receiver but has a generally more 'fragile' look with a long finned barrel and a spindly bipod. In the British Army tests that finally selected the 'Bren', the Vickers-Berthier tended to overheat and its fate was finally sealed when the gun jammed with a live round jammed in the overheated breech and exploded. As war progressed, complications with spare parts and training, coupled with the constant interchange between Indian and British Army units saw the Vickers-Berthier gradually phased out in favour of the 'Bren'. The Vickers-Berthier was manufactured at the Royal Ordnance Factory, Ishapore.

GERMANY

In some ways the impositions of the Treaty of Versailles and the strict limits put on the type and quantity of weapons the Germans could possess worked in their favour for, with no enormous stockpiles of expensive World War 1 machine guns to fall back on, they were forced to build from scratch and were therefore able to be more innovative and to experiment more freely than other nations. This freedom of thought extended not only to weapons design but also to the tactics of warfare and the role of the machine gun.

Where the British and French considered the rifle-carrying soldier to be the basic infantry unit and the role of the machine gun to provide backup, the Germans completely reversed this philosophy, making the machine gun the basic squad weapon, supported by riflemen. In the light of this philosophy, the German approach to machine gun design was different to other nations, for they could see no need for separate medium and light machine guns. In the type of warfare the German planned the machine gun would need to be able to perform both roles; that is light enough to accompany riflemen anywhere yet capable of laying down heavy defensive fire when appropriate. In other words they wanted a General Purpose Machine Gun, although this term would not be adopted officially for another generation.

The first gun to result from this philosophy was designed by Louis Strange of Rheinmetall and built by the company's Swiss subsidiary Solothurn to overcome the Versailles Treaty. Called the 'MG30', it was a very advanced and slender weapon which employed recoil of the barrel to force back the bolt which was unlocked by rollers running in cam tracks in the receiver. The stock was in a straight line with the remainder of the gun which made it very easy to control and incorporated an ingenious quick change barrel. Ammunition was fed from a side-mounted box. A limited number were supplied to the German Army for trial and, after a short while, requests for modification were received back and sample guns passed to Mauser for modification.

Above:
The Germans began the war with what was, in reality, the first 'general purpose machine gun' the 'MG34'.

The gun which Mauser returned bore little resemblance to the one they had received but became one of the most widely used machine guns of the war.

● **MG34** The weapon as revised by Mauser was a beautifully machined tribute to the gunmaker's art. The method of operation had been modified from that of the original 'MG30' and the barrel recoil was now augmented by propellant gases trapped in a muzzle cone and deflected backwards. The bolt locking system was also modified so that only the bolt head revolved and was now locked by an interrupted thread. Also gone was the side mounted box magazine, replaced by a belt feed which, with a quick change of the top cover, could be made to take a 75-round saddle drum magazine to avoid the problems of dangling ammunition belts during movement. Barrel changing was further simplified by hinging the body at the rear of the barrel casing, allowing the body to be swung sideways and the barrel withdrawn. Selective fire was achieved by a 'rocking' trigger assembly where pressure on the top segment achieved semi-automatic fire and pressure on the bottom segment, bursts.

If the 'MG34' had a fault it was that it was too well built; for its fine machining and fitting made mass production impossible and the close tolerances of its working parts made it susceptible to dirt and debris. Despite these failings, five factories worked flat out throughout the war manufacturing nothing but 'MG34s' and the gun appeared on all the battlefields of the German Army. In addition to use on its own integral bipod, the 'MG34' could also be used with a complex tripod which, while light and portable, offered a wide range of adjustment and the facility to fix a dial sight for indirect fire.

● **MG34/41** Intended as a simplification of the original 'MG34' design to speed production. The selective fire facility was removed so that the gun was fully automatic only. The rocking trigger mechanism replaced by a conventional one piece trigger. Internally, the bolt was locked by lugs rather than the more complex interrupted thread and the option of using the drum magazine abandoned. The barrel was also shortened to reduce the recoiling mass of the weapon and increase the rate of fire. Full replacement of the 'MG34' with this variant was prevented by the introduction of the 'MG42'. Main manufacturers of the 'MG34' were Mauserwerke AG, Berlin, Steyr-Daimler-Puch AG, Austria and Waffenfabrik Brunn (Brno) Czechoslovakia.

● **MG42** As the production of the 'MG34' was never going to reach the levels necessary to equip the expanding German Army, design of an alternative was begun. This time the work was not left entirely to gunmakers for experts from the specialist metal stamping and pressing company Johannus Grossfuss, including a Herr Doktor Grunow (a leading expert on mass production techniques), were involved early in the design process to ensure that manufacture would make as much use as possible of sheet metal stampings and pressings, held together by rivets and welding.

The general straight-line layout of the 'MG34' was retained, as was the short recoil operation, augmented by gasses trapped in the muzzle booster assembly. In the 'MG42', steel pressings replaced most of the precisely machined parts that had been a feature of the 'MG34'. The gun assumed a basically square section from the

pressed steel barrel cooling jacket, and receiver to the small wood or plastic stock. The main mechanical difference was in the simplified method of locking employed. Where the 'MG34' had used a rotating bolt and an interrupted thread, the 'MG42' used an idea patented by a Pole, Edward Stecke, in which the head of the bolt carried a pair of small rollers held close to the bolt until the moment of locking when they were forced outwards into matching grooves in an extension of the barrel. The rollers prevented firing pin movement until they were in the fully locked position and, once the locked unit had recoiled sufficiently, cams forced them out of the recesses in the barrel to allow the firing cycle to continue.

The gun fired ammunition from 50-round belt, using a totally new and very effective feed mechanism. Pawls in the top cover of the 'MG42' were operated by a feed arm driven by the reciprocating bolt. These in turn fed the belt through the gun smoothly and reliably and helped the 'MG42'

achieve the astonishingly high rate of fire of 1,200 rounds a minute, the characteristic noise this produced was often said, by those on the receiving end, to sound like 'tearing calico'. The high rate of fire produced considerable vibration and heated the barrels rapidly but these could be changed simply by unlatching the breech end and drawing the whole barrel out through a long slot in the right side of the cooling jacket.

The 'MG42' or 'Spandau' as it was often called by the Allies, had its own folding bipod attached to the front of the cooling jacket but, like the 'MG34,' it could also be used from the Lafette tripod. Because of its high rate of fire, it was also sometimes used in multiple mounts in an anti-aircraft role. The 'MG42' was much more reliable than the 'MG34' in adverse condition and, continued in service to the end of the war. During the war some 750,000 'MG42s' were built by Mauserwerke, Berlin; Johannus Grosfuss, Dohlen; Suhl Waffenfabrik, Suhl; and Steyr-Daimler-Puch in Vienna and many are still in service with armies around the world and so far few of these have become available. When the West German Army was reconstituted, a modernised version, designated 'MG1' and firing the 7.62mm NATO cartridge, was built by Rheinmetall and the Swiss Waffenfabrik, Bern, produced a traditionally machined version, the 'MG51'.

The confusion caused to the Italian rifle development programme by the change of calibre, almost on the eve of war, was as nothing to the confusion it produced in the machine gun field where, during the war, there were at one time or another, machine guns of at least seven different calibres in service with the army. The Italians had been late starters in the development of machine guns, basically, due to procrastination and the inability of those in power to make a decision, and thus fought World War 1 with whatever guns they could buy on the open market - basically Maxim and Vickers guns supplemented by a few inferior domestic designs, rushed into production. In the interwar years, the Italians failed to capitalise on their experiences or those of other nations and so made little progress.

● **FIAT Revelli Model 1914** Originally designed in 1908, by Signor Revelli, and built by the FIAT car company. Despite its very many shortcomings, this gun was still the main infantry support weapon of the Italian infantry at the start of the new war. Outwardly it resembled a Maxim or Vickers gun with its square receiver and cylindrical water jacket but all similarity ended there. The method of operation was delayed blowback, with the barrel and block recoiling together for some half an inch, whereupon the barrel stopped while the block continued. A slight delay was caused by a swinging wedge but even with the underpowered 6.5mm round the action was violent and, to prevent extractors ripping the rims of the cartridge case, an oil pump was fitted to lubricate the cartridges as they loaded. A further peculiarity was the feed mechanism which comprised a metal cage on the left side of the receiver, within which were 10 compartments holding five cartridges each. When each set of five rounds were consumed, the whole unit moved over to expose the next set for firing. An exposed operating rod also oscillated across the top of the receiver during firing, to strike a buffer unit in front of the operating handles. This was a constant hazard to the fingers of the machine gunner and its thin film of oil acted like a magnet to dust and dirt. This and other similar features, contributed to the model's unenviable reputation for jamming. Manufactured by FIAT SpA, Turin.

● **Breda Model 1930** Having originated as a heavy locomotive engineering company, the company of SA Ernesto Breda entered the arms business by way of sub-contracting parts for FIAT machine guns during World War 1, progressing to the full scale production of aircraft guns. In the postwar period, Breda decided to remain in the business and began development of a light machine gun which, as the 'Model 1930', was adopted by the Italian Army. About the only claim to any distinction that can be made for the 'Model 1930' is that it was probably the ugliest machine gun ever produced. There are protrusions everywhere to snag on clothing and equipment, coupled with plenty of holes and slots to collect dust and dirt. The barrel lay in a trough, made necessary to act as a bearing surface because the gun employed a form of delayed blowback in which both barrel and bolt moved backwards. A disadvantage caused by the moving barrel was that the foresight had to be placed on the body of the gun and therefore often did not line up with the barrel. As mentioned with the 'FIAT Revelli', blowback operation presents problems in this type of gun and again, the Italians resorted to forced oiling of the ammunition as it was loaded with all the attendant problems of dust attraction. The 20-round magazine was a permanent fixture on the right side of the gun and was loaded from clips, which lead to a slow rate of fire. A truly awful weapon, without even a carrying handle to ease the gunner's burden, the 'Model 1930' was the Italian's only light machine gun and was thus used extensively. Manufactured by SA Ernesto Breda, Turin.

● **Breda Model 1938** In 1938 the Italians introduced a 7.35mm cartridge and a limited number of the 'Model 1930' machine guns were converted to this cartridge and re-designated. In this variant of the 'Model 1937', the peculiar feed system was replaced by a 20-round detachable box mounted on the top of the gun. A pistol grip also replaced the spade grips of the 'Model 1937'. Manufactured by SA Ernesto Breda, Turin.

● **FIAT Revelli Model 1935** In an attempt to improve the old 'Model 1914' machine gun, the Italians developed the 'Model 1935'. A major improvement in power was achieved by chambering it for the new 8mm cartridge and the complicated and ungainly magazine system was exchanged for a more conventional belt, also the waterjacket was removed . The problem of cartridges sticking in the chamber was overcome by fluting and this allowed the oil pump to be eliminated in many models, although it was later reinstated. Unusually, 'the Model 1935' fired from the closed bolt so that when the trigger was released at the end of a burst, the bolt fed an unfired round into the hot chamber which led to frequent and hazardous 'Cook-offs', with the gun firing itself. With the 'Model 1935,' the Italians managed to achieve an 'improvement' that was actually worse than the gun it was intended to replace. Built by FIAT SpA, Turin.

● **Breda Model 1937** This represents the final attempt of the Italians to produce a satisfactory medium machine gun for its army and was a robust weapon that actually worked reasonably well. Although operated by a gas piston, for some reason the Italians decided not to follow other nations and incorporate a means of gradual opening for the bolt, to ease the cartridge case from the chamber. The action was therefore violent and so back came the pump and oiled cartridges.

The feed system was an oddity, taking rounds from flat tray magazines fed from the side, firing them and then neatly replacing the empty cases in the trays. While this made for tidy machine gun emplacements, achieving it involved lots of mechanical effort and ingenuity and the need to strip empty cases from the trays manually before they could be reloaded.

JAPAN

If the story of Italian machine gun development and acquisition is confused and complex, that of Japan is positively labyrinthine. At the time of the outbreak of war in 1939, Japan had only relatively recently ceased to be an empire of almost medieval feudal states and the remnants of this feudal, diverse thinking led to a large variety of weapons, united only by the common factor that none of them were even remotely satisfactory. The Japanese process of weapons development and procurement demonstrated total lack of any co-ordination, with Army, Navy and Air Force each developing their own weapons totally inde-pendently and with what seems to be an almost defiant refusal to recognise progress made in other parts of the world or to follow logical courses of development. Typical of the peculiar Japanese approach is their decision, in 1932, to change their standard 6.5mm rifle cartridge for something more powerful. In essence, the 7.7mm chosen was a close copy of the British 0.303in round whi.h was quite a logical choice. The Japanese then confused the matter, and presented themselves with enormous supply problems, by producing, in addition to the rimmed version, rimless and semi-rimmed variants.

In fact, the Japanese had been one of the first countries to appreciate the value of the machine gun and, as early as 1902, had purchased rights to manufacture the French Hotchkiss machine gun. From that point on, Japanese development revolved largely around variations on this theme, many of their guns bearing more than a passing

Below:
Japanese troops bring their antiquated Type 92 machine guns into action. *IWM/STT3306*

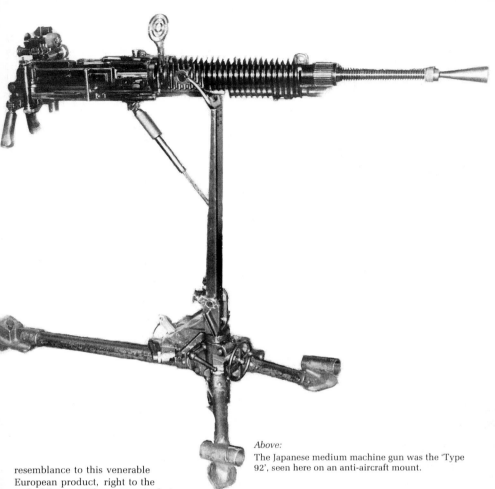

Above:
The Japanese medium machine gun was the 'Type 92', seen here on an anti-aircraft mount.

resemblance to this venerable European product, right to the end of the war. There were, in fact, a very large number of machine guns — some ancient domestic models, others crudely copied from foreign design — in limited service with the Imperial Army during the war. Those listed below are the ones generally accepted as standard issue weapons.

● **Taisho 11** Introduced in 1922 (the 11th year of the Taisho era), this light machine gun was designed by the foremost name in Japanese weapon development, Col Kirijo Nambu. It used the basic Hotchkiss gas operation, with a gas cylinder under the barrel, housing a piston to drive the mechanism. Locking was achieved by a flap, operated by the gas piston. Its most unusual feature was its feed system where a fixed hopper on the left side of the gun was loaded with six standard Arisaka rifle clips of five rounds. The mechanism of the gun took each clip and stripped the rounds, ejecting both empty cases

and clips to the right. The intention was to gain tactical advantage by allowing the gun to be reloaded quickly by ammunition which could be supplied by any infantryman. However, action was violent, particularly with the standard rifle round and eventually a special weaker round was developed for the 'Taisho 11', throwing away this advantage. Even so, the Japanese had to include an oil reservoir to lubricate the rounds to ease the problem of malfunction.

The gun was given a peculiar appearance by the butt and sights which were offset to the right, with the small of the angular butt made of metal. The barrel was finned for almost its entire length. Its complex mechanism was unreliable in adverse conditions and difficult to manufacture in quantity but the gun remained in service until the end of the war.

● **Type 96** Intended as a replacement for the 'Taisho 11', this gun was introduced in 1936 (the year 2596 in the Japanese calendar) and was really a variant of the previous gun but incorporating some features of the Czechoslovakian 'ZB26', probably culled from examples of the gun captured in Manchuria. The basic action was the same as that of the 'Taisho 11' although the hopper feed was replaced by curved, top-mounted, detachable box magazine and the oiling system was removed. Again, the normal rifle ammunition caused problems and the special weak ammunition had to be used and still needed lubrication, an operation which was achieved by a combination oiler/magazine loading tool. The gun was equipped with a ribbed, quick-change barrel, which also carried a bayonet boss and the offset butt was replaced by a conventional straight-line configuration. Normal sighting was achieved by a drum rear sight cribbed from the 'ZB26' although a low power telescopic sight was often fixed.

● **Type 99** This is a modified version of the 'Type 96' which was introduced in 1939 to take the 7.7mm rimless ammunition. It shared the basic outline of the 'Type 96' but was modified internally to provide slow initial opening of the action, making lubrication of the cartridges unnecessary. A flash hider was also fitted to the muzzle. For some reason a short monopod was fitted to the butt to supplement the normal bipod.

● **Type 92 Medium machine gun** This was a direct descendant of the earlier Hotchkiss machine guns used by the Japanese and clearly

Above:
Basic squad support was provided by the 6.5mm 'Type 96' machine gun.

Top right:
The 'Type 96' was one of the few weapons of this type capable of having a bayonet fitted.

Bottom right:
The weapon was often fitted with the low power telescopic sight shown here.

shows its parentage. A heavy, gas-operated, machine gun, it was chambered to fire the 7.7mm semi-rimmed ammunition. The action was modified by Nambu from the earlier 'Taisho 3' machine gun, the main external differences being the addition of a muzzle flash hider and a pair of pistol grips below the receiver. The gun was enormously heavy — 122lb with tripod — and so the tripod was fitted with sleeves on its front legs to admit poles to be attached, allowing two men to support the gun stretcher-fashion, while a third attached a yoke to the third leg. Ammunition feed was by 30-round trays. All Japanese machine guns were manufactured at the State Arsenals.

Being basically an agricultural nation, with little industrial capacity, during World War 1, the Russians relied on machine guns supplied by her allies, predominantly the Maxim. The version used was the 'Model 1910' which, at more than 162lb, unloaded and without water in the cooling jacket, was considerably heavier than the guns of most of the other warring nations and was thus used on a special wheeled mounting. When the Red Army was reorganised in the 1920s, it was quickly established that a light machine gun was a priority. The first steps were already being taken to produce a more manageable machine gun and these centred on variations of the basic Maxim action but air cooled and lightened wherever possible. These were then fitted with wooden stocks and simple bipods to aid mobility. The guns that were developed in this way were only interim measures however, and the Russians realised that few of the light machine guns of World War 1 were worth a great deal. They therefore started from scratch and set in motion the process of designing a domestic model.

● **Degtyarev DP 1928** Vasily Degtyarev was an employee of the Government Arsenal at Tula and was to become one of the Soviet Union's best gun designers. Degtyarev began work on the country's first original light machine gun in 1921 and it was finally adopted by the Red Army in 1928 as the *Ruchnoi Pulemyot Degtyareva Pakhotnyi* which translates to 'automatic weapon, Degtyarev, infantry' and which is understandably abbreviated to 'DP'. It was a remarkable gun in many ways and set the keynote for future Russian weapons in its inherent simplicity, employing only six moving parts and making no demands on skilled labour or advanced machinery in its manufacture. The loaded 'DP' weighed only 26lb and was a conventional gas operated weapon.

The locking mechanism was unusual; on each side of the bolt was a hinged flap which normally lay in a recess flush with the outer surface of the bolt. When the bolt hit the base of a chambered round it stopped but the gas piston continued, taking with it a slider on which was the firing pin. During the final movement, the firing pin acted as a cam to push the locking flaps out of their recesses, into the walls of the receiver, firmly locking the action. Unlocking was achieved as the firing pin moved backwards. Cartridges were fed from a large, flat, dish magazine, driven by a spring and fitted on top of the receiver. This generally worked well although, because of its large areas of thin metal, was prone to damage if not carefully handled.

Overheating was always a problem with the 'DP' for, although the barrel was removable, this could only be done with difficulty and with a special spanner and no spare barrel was carried. The return spring of the 'DP' was coiled around the piston beneath the barrel but it was found that heat quickly ruined the temper of the spring.

● **Degtyarev DT** The Degtyarev Tankovii, was a version of the 'DP' intended for use in armoured vehicles. The main differences were that the magazine was enlarged to take 60 rounds and the barrel was heavier and even more difficult to change. A telescopic butt and pistol grip were attached and a bipod was carried to allow the gun to be removed from the vehicle for infantry use.

Below:
Like all Soviet weapons the 'DP' was simple to make and easy to use. Readily identified by its large drum magazine.

Above:
The 'SG43' has been claimed by experts to be the most successful air-cooled machine gun ever made. Apart from its feed system it is extremely simple and is often seen on this type of wheeled mount.

● **Degtyarev DPM** The 'M' stands for 'Modified' and is a 'DP', introduced in 1944, incorporating improvements made in the light of battle experience. The return spring was moved to a tube overhanging the rear of the gun which made it difficult for the gunner to hold the gun and a pistol grip was fitted. The bipod was also strengthened and moved from the gas cylinder to the barrel.

● **Goryunov SG 43** While the Maxim 'Model 1910' was still being used to provide supporting fire for the Russian infantry, it was obvious that it just was not suited to warfare as waged in the 1940s and that a replacement medium machine gun that was fast and simple to manufacture was desperately needed. The answer was provided by Peter Maximovitch Goryunov with a modification of a tank machine gun upon which he was working. The gun was belt fed, gas operated and had the simple, robust construction that was vital to Russian industrial capabilities. The mechanism was totally new, although it used a conventional gas piston, the bolt locked towards the end of its forward movement when it was forced sideways by a cam, to engage in a recess in the receiver. On firing, the rearward action of the bolt caused a pair of claws to withdraw a cartridge from the belt and forced into a guide, ready for the bolt to move it into the chamber.

The very heavy and easily changed barrel was air cooled and its bore chrome plated, enabling the Goryunov to fire for long periods without overheating. Many of the guns were mounted on a small wheeled mounting, carrying a protective shield for the gunner. In all, six version of the Goryunov were manufactured, differing only in minor detail. Later versions are most easily identified by long grooves cut along the length of the barrel. The guns could also be used on sledge mounts and tripods. The Goryunov is recognised as one of the best air cooled medium machine guns of the war.

UNITED STATES

If there is a single name to be chosen as synonymous with the story of 20th century American machine gun development, it has to be that of John Moses Browning. John Browning was of good Utah Mormon stock and, from an early age, had developed an interest in firearms, acquiring his first patent in the field by the age of 24. He turned his considerable talent to the problems of machine gun design in 1889, having already produced many workable designs for repeating rifles and shotguns. Unlike Maxim, he believed that the energy of muzzle blast offered most potential for the operation of the machine gun mechanism. His first prototypes used a plate attached in front of the muzzle to capture the exhaust gas and, through a system of levers, to power the reloading cycle. From this premise he moved on to drilling a hole in the gun barrel close to the muzzle to lead gas into a cylinder where it impinged upon a piston which in turn operated the mechanism. This was the first true gas operated gun

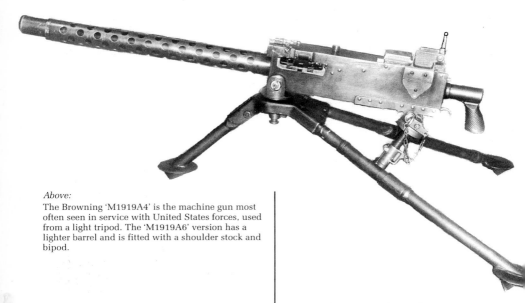

Above:
The Browning 'M1919A4' is the machine gun most often seen in service with United States forces, used from a light tripod. The 'M1919A6' version has a lighter barrel and is fitted with a shoulder stock and bipod.

and was built by the Colt company for the US Navy in 1895 and thus became the United States' first automatic weapon.

Having handed over his gun to Colt, Browning returned to the drawing board and began to reconsider his predilection for gas operation, experimenting with a variety of recoil operated designs. Little interest was shown in his designs until, in 1917, he was asked to demonstrate his new machine gun to the Army. In one test, 20,000 rounds were fired through a gun without stop and then, after a brief halt another 20,000. The inspecting board were sceptical that a production gun could perform so well so Browning fetched another gun and fired it non stop for more than 48min, during which it consumed nearly 30,000 rounds of ammunition. Within weeks orders had been received for 10,000 machine guns and 12,000 of a machine rifle he had also designed. Both designs were used extensively throughout World War 1 and into the next.

● **0.30in-'06 Browning M1917A1** The mechanism of the Browning is operated by recoil of the barrel. After a short distance, a vertical lock is withdrawn by cams in the body of the gun, unlocking the bolt. The final section of the barrel's movement causes a curved steel claw, called the accelerator, to swing back and, because of mechanical gain through leverage, this shoots the bolt back very rapidly against a return spring. This movement of the bolt also drives the belt feed mechanism to strip rounds from the belt and position them for loading. This was the water-cooled machine gun which the Americans used throughout World War 1 and into World War 2 where, with minor modifications it was designated the 'M1917A1'.

● **0.30in-'06 Browning M1919A4** A lighter and more portable version of the 'M1917' with a heavier barrel but with the water jacket removed in favour of a ventilated barrel jacket. The rate of fire was slower and the gun was intended to be used from a tripod. The sights were also slightly modified.

● **0.30in-'06 Browning M1919A6** Intended to improve the tactical capabilities of the machine gun, a bipod, wooden shoulder stock and carrying handle were added in an attempt to transform it into a light machine gun. Retaining the ventilated barrel jacket the gun had a lighter barrel than the 'M1919A4' and thus had a higher rate of fire. These weapons were manufactured by Remington Arms, Winchester Repeating Arms Co; Westinghouse Inc and by Colt Patent Firearms Corp.

● **Browning Automatic Rifle M1918A2** Although officially called the 'Browning Automatic Rifle' (or 'BAR') this weapon is really a machine rifle or light machine gun in its capabilities. John Browning had designed it in 1916 as a 'light' shoulder fired rifle for use in advance across No Man's Land, although at 16lb, few would actually call it light. In fact it falls between being an automatic rifle and a light machine gun — in other words a compromise, but one that worked reasonably well and was liked by those that had to use it.

The mechanism, which was gas operated, used a tipping bolt modelled after one of Browning's successful shotgun designs. The weight constraints of weapon led to the need for light reciprocating parts and this would in turn have led to an unacceptably high cyclic rate without the incorporation of a shock absorber in the return spring assembly. Even so, the violence of the action caused rapid wear and the 'BAR' tended to wear out far quicker than most other similar weapons.

The original version was intended to be fired from hip or shoulder and only later was a barrel-mounted bipod added. The 'M1918A2' which was adopted shortly before World War 2 began, was used as an infantry light support weapon and was fitted with a robust bipod. Its main limitation, for a weapon that would only fire in the fully automatic mode, was that its bottom mounted magazine held only 20 rounds and was thus rapidly exhausted. The weapon also heated rapidly and was therefore any way unsuitable for sustained fire. An interesting feature of this model 'Bar' is the possibility of setting it to either slow or fast rates of fire. Examples were built by Colt Patent Firearms Manufacturing Co, Hartford, Connecticut; Winchester Repeating Arms Co, New Haven Connecticut; the Marlin-Rockwell Corp; IBM; and the New England Small Arms Corp.

Below:
Although officially designated as a rifle, the Browning 'M1918A2' was most often used as a light machine gun by the US forces and large numbers were supplied to Britain for the Home Guard.

Machine Gun Technical Data

Britain & Commonwealth

Type	Operation	Calibre	Weight	Length	Barrel length	Feed	Cooling	Cyclic rate	Muzzle velocity
Vickers M1	Recoil	0.303in	40lb (tripod) 50lb	45.5in	28.5in	250 round belt	water	450 rounds/min	2,450ft/sec
Bren Mk1	Gas, selective	0.303in	(gun) 22.5lb (tripod) 26.5lb	45.5in	25in	30-round box or 100-round drum	air	500 rounds/min	2,440ft/sec
Bren Mk2	Gas, selective	0.303in	(gun) 23lb (tripod) 26.5lb	45.6in	25in	30-round box or 100-round drum	Air	540 rounds/min	2,440ft/sec
Bren Mk3	Gas, selective	0.303in	(gun) 19.3lb	42.6in	22.25in	30-round box	air	480 rounds/min	2,440ft/sec
Bren Mk4	Gas, selective	0.303in	19.14lb	42.9in	22.25in	30-round box	Air	520 rounds/min	2,440ft/sec

Germany

Type	Operation	Calibre	Weight	Length	Barrel length	Feed	Cooling	Cyclic rate	Muzzle velocity
MMG34	Recoil, selective	7.92mm	(gun) 26.5lb (tripod) 42.3lb	48in	24.75in	50-round belt	Air	800-900 rounds/min	2,500ft/sec
MG34/41	Recoil, automatic	7.92mm	(gun) 26lb (tripod) 42.3lb	46in	22in	50-round belt/50-round belt drum/ 75-round saddle drum	Air	800-900 rounds/min	2,440ft/sec
MG42	Recoil, automatic only	7.92mm	(gun) 25.5lb (tripod) 42.3lb	48in	21in	50-round belt/50-round belt drum	Air	1,200 rounds/min	2,480ft/sec

Continued overleaf

Italy

Type	Operation	Calibre	Weight	Length	Barrel length	Feed	Cooling	Cyclic rate	Muzzle velocity
FIAT-Revelli Model 1914	Delayed blowback	6.5mm	(gun) 37.5lb (tripod) 49.5lb	46.5in	25.75in	50-round	Water	400 rounds/min	2,080ft/sec
Breda Model 1930*	Delayed blowback/ selective fire (automatic only)	6.5mm	(gun) 22.75lbs (tripod) 49.5lb	48.5in	20.5in	20-round fixed 'mousetrap' magazine	Air	475 rounds/min	2,063ft/sec
	* Details of the 7.35mm Model 1938 are virtually the same.								
FIAT-Revelli Model 1935	Delayed blowback,	8mm	(gun) 40lb (tripod) 41.5lbs	49.75in	25.75in	300-round belt	Air	500 rounds/min	2,600ft/sec
Breda Model 1937	Gas automatic selective fire only	8mm	(gun) 42.8lb (tripod) 41.5lb	50in	25in	20-round strip	Air	450 rounds/min	2,600ft/sec

Japan

Type	Operation	Calibre	Weight	Length	Barrel length	Feed	Cooling	Cyclic rate	Muzzle velocity
Taisho 11	Gas — automatic only	6.5mm	(gun) 22.5lb	43.5in	19in	30-round	Air	500 rounds/min	2,300ft/sec
Type 96	Gas — Mautomatic only	6.5mm	(gun) 20lb	41.5in	21.7in	30-round box/ Hopper — detachable	Air	550 rounds/min	2,400ft/sec
Type 99	Gas — automatic	7.7mm	(gun) 23lb 122lb with tripod	46.75in	21.5in	30-round box	Air	850 rounds/min	2,350ft/sec

Type 92 Medium Machine Gun

	Operation	Calibre	Weight	Length	Barrel length	Feed	Cooling	Cyclic rate	Muzzle velocity
	Gas — automatic	7.7mm	(gun) 23lb 122lbs with tripod	45.5in	29in	30-rd strip	Air	450 - 500 rounds/min	2,400ft/sec

Soviet Union

Type	Operation	Calibre	Weight	Length	Barrel length	Feed	Cooling	Cyclic rate	Muzzle Velocity
Degtyarev DP 1928	Gas — automatic only	7.62mm	(gun) 26.23lb	50in	23.8in	47-round drum	Air	500-600 rounds/min	2,756ft/sec
Degtyarev DP	Gas — automatic only	7.62mm	26.9lb	50in	23.8in	47-round drum	Air	500-600 rounds/min	2,756ft/sec
Degtyarev DT	Gas — Mautomatic only	7.62mm	(gun) 27.91lb (mount) 59.3lb	46.46in	23.5in	60-round Drum	Air	600 rounds/min	2,756ft/sec
Goryunov SG 43	Gas — automatic only	7.62mm	(gun) 30.42lb (mount) 59.3lb	44in	28.3in	60-round Belt	Air	600-700 rounds/min	2,832ft/sec

United States

Type	Operation	Calibre	Weight	Length	Barrel length	Feed	Cooling	Cyclic rate	Muzzle velocity
Browning M1917A1	Recoil — automatic only	0.30in — '06	(gun) 32.6lb (tripod) 53.15lb	38.5in	24in	250-round belt, fabric or disintegrating link	Water	450 - 600 rounds/min	2,800ft/sec
Browning M1919A4	Recoil — automatic only	0.30in — '06	(gun) 31lb (tripod) 53.15lb	41in	24in	250-round belt, fabric or disintegrating link	Air	400-550 rounds/min	2,800ft/sec
Browning M1919A6	Gas — Mautomatic only	0.30in — '06	(gun) 32.5lb	53in	24in	250-round belt, fabric or disintegrating link	Air	400-500 rounds/min	2,800ft/sec
Browning Automatic Rifle M1918A2	Gas — automatic only	0.30in — '06	(gun) 9.4lb	47.8in	24in	20-round detachable box magazine	Air	300-450 rounds/ min (slow) 500-650 rounds/min (fast)	2805ft/sec

6

Submachine Guns

Major conflicts usually evolve an infantry weapon that is uniquely its own; the Boer War promoted the use of the bolt action magazine rifle while in World War 1, no combatant nation could afford to be without the machine gun. For World War 2, the equivalent weapon is probably the submachine gun. A common feature it shares with the other weapons mentioned as examples is that none were new when their respective wars began but were brought into prominence as developing tactics and usage made particular features invaluable. In the case of the submachine gun, its origins and first usage were some 20 years before to help break the stalemate of European trench warfare.

The submachine gun was originally conceived as a light, mobile, hand-held weapon to be used to provide concentrated firepower for close quarter combat such as trench clearance, street fighting or on patrols. In these situations, the enemy might put in only a fleeting appearance at close range and from any direction and the soldier operating in these conditions required a light, handy and controllable weapon that would allow him to maximise his chances of hitting his target by providing a short burst of shots at a single squeeze of the trigger.

The problem with available machine guns was that they were both heavy and cumbersome and adapting them was impractical as their powerful rifle calibre ammunition made them difficult to control without sufficient bulk to control and absorb recoil. The solution, first arrived at by the Italians, was to develop a weapon firing standard pistol ammunition in the fully automatic mode. The Italian weapon was the rather odd and ungainly 'Vilar Perosa' which, although designed to fire pistol calibre ammunition, was used more in the role of a light machine gun. It was usually mounted either on a bipod or on a wooden tray carried on straps over the shoulders and consisted of two weapons built and assembled side by side but fitted with a pair of brass spade grips. A pair of box magazines on top of the weapon supplied ammunition.

At almost the same time, the Germans were working on a weapon that was to be much more like the type of submachine guns we see today. In

Right:
The weapon on World War 2, the submachine gun evolved enormously during the war. Pictured, from left to right, are the British 'Sten' MkII, the Soviet 'PPS42' and the American 'Thompson'.

Below:
Paratroopers advance through Oosterbeek in Holland; two are carrying the Mk V 'Sten' gun. *IWM/BO1121*

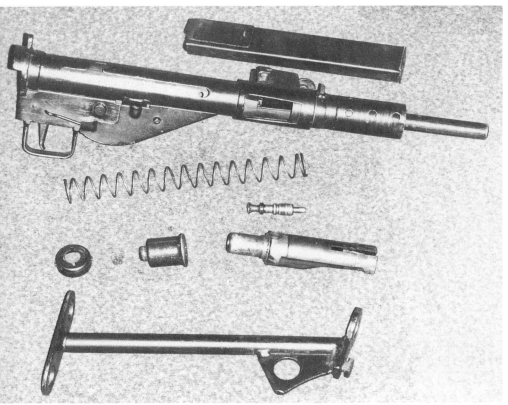

108

an attempt to break the stalemate on the Western Front, the German army had begun to experiment with *Stosstruppen* or 'Storm Troops', highly mobile groups of soldiers, lightly equipped but well armed, intended to infiltrate defences in small groups to open a path for massed infantry. To succeed, such troops needed overwhelming firepower and to meet this requirement the 'Bergmann Musquette' or 'Kugelspritze' (bullet squirter) was designed by Hugo Schmeisser.

Construction of his weapon, which was eventually given the official designation 'Maschinen Pistole 1918' or 'MP18' was comparatively simple, basically just a tubular receiver housing a heavy bolt and return spring. The barrel was covered by a perforated jacket/handguard and the mechanical components, together with a simple trigger mechanism, were seated in a wooden rifle-style stock. 9mm Parabellum ammunition was fed from the 32 round 'snail' magazine developed for the 'P '08 Luger' pistol.

In operation, the bolt was pulled back by a cocking handle, against the pressure of the weapon's recoil spring, until it was held by the sear of the trigger mechanism. When the trigger was pulled, the bolt was released to pick up a cartridge from the magazine, which was chambered and then fired by the firing pin housed in the bolt. The exploding charge forced the bolt back against the spring, allowing the extractor to pull the empty case from the chamber for ejection. If the trigger was kept depressed, then the cycle continued; if trigger pressure was released, the bolt was held back at the end of its travel by the sear. The 'Bergmann' was to set the pattern for submachine guns for many years to come.

When used in the Luger pistol, the relatively powerful 9mm Parabellum cartridge required a complex and sensitive 'toggle link' mechanism to lock barrel and breech together until chamber pressure had reached a safe level.

Schmeisser wished to keep his weapon simple. However, a plain unlocked breech in a blowback weapon of this type would have required an extremely heavy bolt and an uncomfortably strong recoil spring. The answer was to use 'differential locking' or 'advanced primer ignition'. In the 'Bergmann', the recoil spring acted directly on a firing pin carrier, ensuring the pin always stayed proud from the face of the bolt. Also on

Top left:

Internally they changed too. The complexity of the Thompson '1928A' ...

Bottom left:

... gave way to the stark and practical simplicity of the minimalist 'Sten'.

the face of the bolt was a broad and strong sprung extractor claw, designed to clamp over the base of the cartridge to withdraw it from the breech on recoil. In operation, when the bolt picked up a cartridge from the magazine and forced it into the chamber, it was the extractor claw which actually pushed it into the breech, keeping the face of the bolt and the firing pin clear of the primer.

Eventually a point of sufficient resistance was met to overcome the spring pressure of the extractor claw, allowing it to snap over the base of the cartridge into the extractor groove.

The firing pin was then free to strike the primer and fire the cartridge. It did this while the bolt was still moving forward and thus the explosion of the charge had to halt and overcome the forward momentum of the bolt, before it could reverse its direction. This delay was sufficient to allow chamber pressure to drop before the bolt began its rearward movement under recoil and made it possible to keep the working parts of the mechanism relatively light.

Having developed and introduced a light and mobile automatic weapon, the German army immediately robbed it of some of its tactical advantages by adopting an operational philosophy that dictated that each submachine would have a crew of two people; one to act as principal gunner and another who would act as ammunition carrier. In addition, each pair of gun teams were supplied with a handcart to transport ammunition, some of which was boxed and some loaded into spare magazines. Even so, some 35,000 'Bergmanns' were issued before World War 1 ended and, while the weapon did make an impact on trench warfare, its advent was too late to affect the overall course of the war.

Once the armistice was signed, attitudes towards the submachine gun became mixed. In the majority of military authorities, the opinion was held that this type of weapon had been spawned by the particular conditions of warfare on the Western Front, which were unlikely to be repeated in any future conflict. As a consequence, the Treaty of Versailles prohibited the peace-time German army use of the submachine gun (although it was allowed for some police units). Of those armies that were free to adopt the weapon, few chose to do so and the submachine gun lapsed into virtual obscurity until the beginning of the Spanish Civil War.

It was this war that saw the first sizeable use of submachine guns and the first realisation of their true tactical value. The two man team and supporting ammunition carts disappeared and the guns were extensively used to provide the individual infantryman greater firepower without affecting mobility. The war also saw the birth of the realisation that the submachine gun did not need to be carefully made and expensively machined and that standards of finish and craftsmanship could be lowered without affecting the safety or performance of the weapon. It also demonstrated the advantage that a simple automatic weapon offered in the hands of untrained or partly trained troops.

Notably, as a consequence of the experience of the Spanish Civil War, some nations, mainly Germany, Russia and Italy, began designing and mass producing submachine guns and were thus able to have them in the hands of their troops, in quantity, before the outbreak of World War 2, not so Britain and, to some extent, the United States.

BRITAIN AND COMMONWEALTH

The British just did not appreciate the potential of the submachine gun and, in fact, it was often referred to in defence circles as 'a gangster gun' that had no place in the armoury of a civilised nation and, typically, is dismissed in one paragraph in the Text Book of Small Arms 1929. This attitude originated at the tail end of World War 1. A captured 'Bergmann' was tested and the General Headquarters in France was asked whether they saw a requirement for this type of weapon at the Front. The reply concluded: 'A really penetrating bullet is necessary to ensure that the enemy's problems in regard to penetration shall remain difficult. It follows therefore that no weapon of the pistol nature can ever replace the rifle as the infantry's main arm and no gun resembling this particular German weapon is required in the British Army'. This response ensured the British Army would eventually be denied a vital weapon type at a time of critical need.

When war broke out and the nature of the German threat became apparent, the British were forced into a drastic rethink and an almost panic stricken programme to develop and produce a home-bred submachine gun. This was prompted by an urgent request, made at the end of 1939, by

Below:
The Lanchester was Britain's first submachine gun and was originally intended as a stopgap for the army although most eventually ended up in naval service.

the British Expeditionary Force for an immediate supply of 'machine carbine or gangster weapons'. At this stage options were strictly limited, and in reality the only true short term solution was to buy guns abroad. This resulted in purchase, as an interim measure, of quantities of the American Thompson submachine gun which is detailed later. This was an expensive option as each gun cost roughly $50 plus another $5-10 for the Cutts Compensator, which was needed to help reduce muzzle climb. As this outlay represented something like 1% of the cost of a Spitfire fighter aircraft, a cheaper, home-bred weapon was urgently needed.

● **LANCHESTER** Filling the gap in the armoury was by no means a simple task. In January 1940, a captured example of the 'Schmeisser' gun was tested and deemed to be 'reliable and accurate'. In fact this was the 'Bergmann MP28', a development of the World War 1 'MP18', which had been altered to allow the use of straight 'box' magazines rather than the complicated 'snail' drum magazine of its predecessor and which had the ability to fire single shots as well as fully automatic bursts.

It was therefore decided to copy the 'MP28' in every detail. Sterling Armaments was given the task of building a prototype and producing the weapon under the direction of Mr George Herbert Lanchester and, by the end of November, the first limited endurance trial was held using the fourth prototype built by Sterling. This fired more than 5,000 rounds with only 26 stoppages, most of which were caused by faulty ammunition. In fact, the Lanchester never did reach the army, although it did equip the Royal Navy. The Lanchester was a typical Navy weapon; solid, heavy and reliable and, with British industry not, at the time of construction, working on a war footing, the workmanship and finish (particularly of the early weapons) was of high quality.

The Lanchester Mk1 was a total copy of the 'MP28', and was fitted with a walnut rifle-style stock which showed distinct Lee Enfield parentage, right down to the brass butt plate. The housing for the 50-round box magazine was also of brass and the receiver mounted a rifle-style tangent sight. The perforated barrel jacket was fitted with a standard boss to take the Lee Enfield 1907 bayonet and the weapon was provided with a change lever situated just ahead of the trigger guard.

Above:
An exact copy of the World War 1 German
'Bergmann', the Lanchester was a reproduced
'antique'.

The Lanchester Mk1* differed only in detail.
The change lever was removed so that the gun
could only fire fully automatic and the complicated and unnecessary tangent rear sight was
replaced with a simplified 'flip over' two position
rear sight. Many Mk1s were converted to Mk1*s
and so examples can be found minus change
lever but with tangent sight.

Although an historically interesting weapon,
the 'Lanchester' actually saw little real service in
battle, being used mainly by boarding parties and
for the occasional commando raids, although the
weapon did remain in Navy armouries through
the Korean War and on into the 1960s, before
eventually being replaced by its direct descendant the 'Sterling L2A3'.

● **STEN** Even while the 'Lanchester' was entering
production, work was being conducted to
develop a cheap and simple new design that
could be manufactured in the quantities
demanded by the armies being raised in the Commonwealth. In the early days of 1941 'NOT 40/1'
arrived on the scene; a design originated by
Maj R. V. Shepard, a director of the Birmingham

Small Arms Co (BSA) and Mr H. J. Turpin, the
company's principal designer. By the end of January, the prototype had been fully tested and
accepted with a decision that first production
models were to be received from factories in
June. The weapon was to be given the official
designation 'Carbine, machine, Sten' and was to
prove to be the archetypal submachine gun that
would set the standards of finish and manufacture for years to come. The name derived from
the initial letters of the surnames of the two
designers, coupled with 'EN' from the location of
the Royal Small Arms Factory at Enfield.

The key factors in the design philosophy were
ease of manufacture and simplicity of operation.
In fact the designers made the most of modern
production techniques and came up with a
design that made maximum use of general, rather
than small arms, manufacturing technology. This
brought with it the benefit that many parts could
be sub-contracted for manufacture. The result
was extremely crude and earned it titles such as
'the Plumber's delight' and the 'Stench Gun'
among its more polite nicknames.

The Sten worked on a straight blow back principle, employing a heavy bolt with a fixed firing
pin and a large coil return spring. The body was a
simple tubular affair and, for ease of manufacture, frills were kept to a minimum.

● **MkI Sten** The 'MkI Sten' was relatively elabo-
rate in comparison with marks that were to follow. It incorporated a wooden fore-end and a

wooden bracer in the small of the steel skeleton butt. The fore-end also had a crude tubular steel pistol grip which could be folded away underneath the barrel when not in use. The barrel was enclosed by a ventilated tubular casing and carried a large conical flash hider. The gun also boasted a change button to allow selection of single shot or automatic fire. The process of further simplification began almost immediately and, before the end of 1941, the 'Sten MkI*' appeared and in this model the woodwork, the flash eliminator and the folding pistol grip were removed.

Both models had a fixed, six groove rifled barrel. In total some 100,000 'MkI' and 'MkI* Stens' were manufactured.

● **MkII Sten** This model is often described as the ugliest and nastiest weapon ever employed by the British army and this is a perfectly apt description for it took the principles of cheap mass production almost to the extreme. It was however, also an effective and deadly weapon that could kill people as efficiently as even the most expensive submachine gun.

All the woodwork used on previous models disappeared, as did the barrel jacket. Wood was replaced by a sheet metal covering for the trigger mechanism. A removable, two groove barrel, retained by a perforated nut which could be used as a hand grip, was fitted and the magazine housing could be rotated through 90° to act as a dust cover for the ejection port in bad conditions. Standards of finish were poor with blobs of welding metal on the seams left unsmoothed. The weapon was painted rather than blued or parkerised. In most models, the butt was a simple tube with welded cross piece.

A Second Pattern 'MkII Sten' was introduced to reduce further the material used in production and many of these were produced in Canada by the Long Branch Arsenal. They are characterised by a better external finish than their British counterparts and a stronger skeleton butt. This model could also be fitted with a bayonet.

● **MkIII Sten** Yet further simplification was exhibited by this mark, first produced in 1943 by the toy maker Lines Brothers. The entire body was a tube of sheet metal, riveted and with a long weld running its full length. The non removable

barrel projected only an inch from the body. The magazine housing was rigidly fixed and the stock reverted to the single tube and crosspiece of the early 'MkIIs'. This model was manufactured in Britain and Canada.

● **MkV Sten** The final wartime issue version of the 'Sten', the 'MkV' was generally considered to be one of the best submachine guns in service during World War 2. It was generally better made than its predecessors, with a wooden butt and pistol grip and early versions carried an additional vertical pistol grip at the fore-end. The front sight was taken in toto from the 'No4' rifle. This mark was introduced in 1944, was used extensively by the airborne forces at Arnhem and remained in service into the 1950s.

● **MkII(S)** Commando and other special forces had a particular requirement for a silenced weapon with which to deal with opposition while retaining stealth. The 'MkII(S)' was developed for this purpose. The weapon's barrel was drilled to bleed off gases into a jacket extending well beyond the muzzle. Metal cups surrounded the barrel and continued forward of it. As gases emerged from the barrel drillings, their energy was dissipated by the cups. At the end of the jacket the bullet penetrated a rubber sealing plug that prevented gases following it. The jacket was itself perforated and in turn surrounded by a metal sleeve. The gases heated the outer case and so the assembly was wrapped in a thick canvas cover to protect the user's hands.

Users were advised only to fire single shots as bursts quickly burnt out the silencer. As the drilling of the barrel reduced the power of the cartridge, a lightened bolt was used and the final two coils of the return spring were cut off to reduce its strength. It was an effective weapon to 100yd with only the sound of the bolt moving audible fifty feet from the muzzle.

● **OWEN** With Britain hard pressed to fulfil its own requirements for submachine guns, it was in no position to supply such weapons in quantity to the Australians who therefore began their own development programme. Lt Evelyn Owen produced a prototype submachine gun in late 1940 that was immediately accepted and which remained in service until 1962. The 'Owen' was superficially similar in outline to the 'Sten', although it had a rather strange and fragile look. Its most prominent feature was a top-mounted magazine, an arrangement that was less awkward than may at first be apparent as gravity assistance to the internal spring made for reliable feed and few stoppages and, with the centre of gravity almost directly over the pistol grip, it was possi-

Left:
For clandestine operations the Mk2S, or silenced 'Sten', was developed.

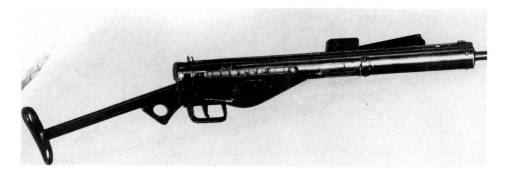

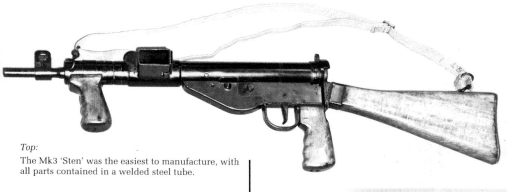

Top:
The Mk3 'Sten' was the easiest to manufacture, with all parts contained in a welded steel tube.

Above:
The final version of the 'Sten', the Mk5, was introduced in time for the Battle of Arnhem. It had a wooden butt and the foresight of the No4 rifle, allowing it to accept a bayonet.

ble to fire one-handed. This arrangement made offset sights a necessity but was a useful feature in the Far Eastern campaigns, where the magazine was less likely to snag jungle foliage. In fact, although the 'Owen' bore a passing resemblance to the 'Sten', it was a much better made weapon, built by traditional gunmaking methods and relying heavily on precision machining processes, which limited its production to low numbers. Unlike the 'Sten', the 'Owen' was very popular with troops. The gun was manufactured by Lysaght Pty, Port Kemble, New South Wales. Models built after 1943 are characterised by a painted jungle camouflage finish.

Right:
The Australians developed their own submachine guns. The first was the Owen.

Left:
The Owen was followed by the less popular Austen.
This is an Austen Mk2, which made use of
aluminium castings.

● **AUSTEN** Although the 'Owen' was a good, efficient weapon, it did not lend itself to mass production. Therefore, shortly after the first 'Sten' guns appeared in Australia a home-brewed variant, known as the 'Australian Sten' or 'Austen', was evolved. The Australians copied the body, barrel and trigger mechanism of the 'MkII Sten' and married this to a bolt mechanism cribbed from the German 'MP38' which had a separate firing pin and a telescopic cover for the recoil spring. The Australians also adapted and improved the German folding stock. The 'Austen' was unpopular with troops, most of whom preferred the 'Owen' and, although specifically designed for quantity production, only some 20,00 were actually built. The 'Austen' was manufactured by two companies; Diecasters Ltd and W. J. Carmichael & Co Ltd, both based in Melbourne.

GERMANY

The strict provisions of the Treaty of Versailles would seem to have put paid to the design and manufacture of submachine guns as their numbers and use were limited only to the requirements of the police. However, the Germans had established links with manufacturing companies in other countries that allowed them to circumvent the Treaty conditions. Typical was that established by Metallwaren und Maschinenfabrik, based in Dusseldorf, later to become better known as Rheinmetall. The company bought stock in the Oesterreichishe Waffenfabrik Gesselschaft at Steyr in Austria and established a subsidiary company, Waffenfabrik Solothurn in Switzerland. This arrangement enabled the German parent company to design weapons, which did not contravene the Treaty, while the prohibited development engineering was conducted in Austria and production in Switzerland.

Such arrangements enabled Germany to develop, test and manufacture guns despite the Treaty and, at the outbreak of war, she was able to field an army equipped with some of the best and most advanced small arms available. The process that led up to this situation was one of gradual evolution rather than quantum leaps.

Logically, the Germans began the process of evolution from the 'MP18' which had served them so well in World War 1. The first derivative was the 'MP28'. Practical experience with the 'MP18' had indicated that some modifications were required. The most important was the incorporation of a new magazine housing to hold a simple box magazine rather than the complicated, clockwork-powered 'snail' magazine. Second was the addition of a stud above the trigger which allowed single shots or fully automatic fire. The general appearance of the weapon did not change. Small quantities of the 'MP28', within the limits set by the Treaty of Versailles, were manufactured in the Haenel factory at Suhl in Germany but far more were manufactured under licence at Herstal in Belgium.

Although production of the 'MP28' was discontinued before the outbreak of war, the weapon did equip some special units of the German army, often those formed for policing operations and whose members had experience of the gun in the interwar period. Other weapons were developed, similar to the 'MP28', which in having a conventional wooden stock and a box magazine horizontal to the barrel relied heavily on traditional gunmaking methods.

During the 1930s, the Germans began to develop tactics for its armour. Until then, the tank had been considered a separate entity but the Germans began to look at it as a part of a balanced, independent force — the Panzer Division — that would have a key role in fast moving 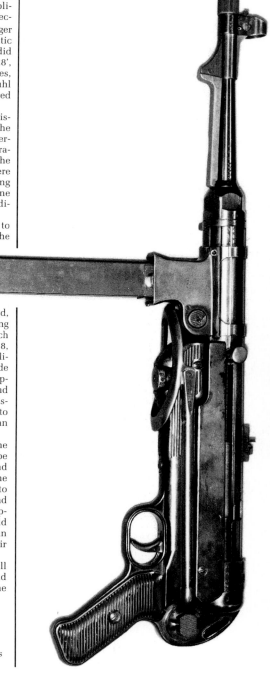 *Blitzkrieg* or 'lightning warfare'. At its spearhead, the Panzer Division would have hard-hitting motorised infantry battalions, operating much along the lines of the Storm Troopers of 1918, and these would need to be equipped with a reliable, easily produced submachine gun to provide the necessary automatic fire. The task of developing the new weapon was given to Ermawerke and was to lead to one of the world's most recognisable submachine guns and one that was to become the virtual trademark of the German army.

● **MP38** This weapon and its successor the 'MP40' were the first submachine guns to be taken into large scale use by major power, and are often and quite erroneously called 'The Schmeisser'. This is purely a name ascribed to them by the Allies for Hugo Schmeisser had absolutely no hand in their design and development, this being entirely the work of Berthold Giepel of Ermawerke. In fact, to the German army, the guns were only ever known by their numerical designation.

With his design, Giepel threw away almost all previous ideas. Gone were the wooden stock and the perforated barrel jacket that had been the

Right:
The main submachine gun of the German Army was the so-called 'Schmeisser' — the 'MP40'.

style to date. Instead, all possible steps were taken to simplify and cheapen production. The body was made from steel tubing with 'fullering' — longitudinal grooves machined along its full length — to reduce weight while increasing stiffness. The cover for the mechanism was cast from aluminium and then anodised and the fore-end and pistol grip were manufactured from a plastic resin. The traditional wooden stock was replaced by an oval metal ring attached to two metal rods that could be folded down to lie forward under the body. The magazine also hung vertically below the body, from a housing that could be used as a forward grip.

Internally, the bolt contained a sprung firing pin and was pushed forward by a return spring housed in a telescoping metal sleeve. Despite the steps taken to simplify construction, the 'MP38' was still costly to manufacture, requiring extensive milling and machining. It was built in only limited numbers by Erma.

● **MP40** This weapon, although substantially similar to the 'MP38', was introduced specifically to take most advantage of simple manufacturing processes. Very little high grade steel or aluminium was used in the gun. The body and as much as possible of the mechanism was formed by stamping from sheet metal which was then either brazed or spot welded and much of this work was sub-contracted to minor firms.

The 'MP38', in common with many blowback submachine guns, had a major safety failing. Its safety catch was merely a slot into which the one piece cocking handle could be fitted when the weapon was cocked. This did not allow the weapon to be carried safely in the uncocked condition, with bolt forward on an empty chamber. If the gun was dropped, or jarred heavily, the bolt could bounce back far enough to pick up a round from the magazine, chamber and fire it. To remedy this, the 'MP40' was given a two piece cocking handle and a further slot was cut in the forward end of the bolt receiver track to lock the bolt. Between 1940 and 1944, more than 1,000,000 'MP40s' were produced by Erma, Haenel and Steyr. Early 'MP40s' had a magazine housing with a smooth surface and these are comparatively rare. More common is the 'MP40/1', which has a ribbed magazine housing. A rare variation is the 'MP40/2' with a magazine housing carrying two magazines side-by-side, allowing the spare magazine to be slid across when the first was emptied.

Above:
The much more widely used 'MP40' was much quicker and cheaper to manufacture. It was largely made from stampings and with a redesigned cocking handle.

ITALY

Although the Italian army had been among the leaders in the use of submachine guns, they had, to some extent, allowed this lead to slip in the years between the wars. The strange 'Villar Perosa', which was a submachine gun only in its method of operation and use of a pistol cartridge had, towards the end of World War 1, been superseded by more practical and more typical submachine guns. In fact Villar Perosa produced a single barrelled version, with a conventional rifle stock that remained in service through into the North African campaigns. Beretta also produced several intermediate designs with varying degrees of success.

In the late 1930s, firm interest in a modern submachine gun again arose in the Italian army and Beretta was asked to develop some designs and the task was passed on to Tullio Marengoni, Beretta's designer, who had been toying with submachine gun designs since 1918.

● **Beretta M1938A** The 'M1938A' was derived by Marengoni from his design for a self-loading 9mm Parabellum carbine intended for limited police use. This parentage shows in the general outline of the submachine gun as in appearance it is more like a short rifle than a typical submachine gun. It had a one piece, conventionally shaped wooden stock, topped by a long cylindrical receiver and barrel assembly.

The overall length of all members of this family is a fraction over three feet. The early guns
were true products of the gun makers' art. Even
without a magazine, which came in 10, 20 and
40-round versions, the gun weighed nine and a
half pounds. They were fitted with rifle sights,
graduated from 100 to 500m. The sighting
arrangement may seem optimistic for a 9mm but
a special high velocity round was developed and
used at one stage in an attempt to match the
accuracy and range capability of the gun. The
weapon was capable of full or semi-automatic
fire but, in place of a conventional change lever,
the Beretta was fitted with two triggers, situated
in the trigger guard in tandem. The front trigger
was for automatic fire while the rear trigger
allowed semi-automatic single shots. Another
departure from most submachine gun designs is
the siting of the ejection port on the left side of
the gun.

The very early models can be distinguished by
the elongated slots cut into the barrel jacket as
ventilation and by the compensator which takes
the form of a single large hole cut in the top of
the muzzle. A folding knife bayonet was also fitted. The elongated cooling slots were used on
only a very few examples as was the folding bayonet, most versions being provided with circular
vents and a boss and slot mounting arrangement
for a conventional separate bayonet.

The method of operation was simple blow
back, the gun firing from the open breech and the
bolt cocked by a handle on the right of the gun,
attached to a cover slide which engaged the bolt
only for cocking. The handle was pushed forward
after cocking and was held forward by a catch so
that it did not reciprocate with the bolt. The
cover slide also acted to exclude dust.

For a submachine gun the method of firing
was complex and again added to the cost. When
the bolt was forward, a cam made contact with
the ejector stud and was revolved to drive the firing pin forward. The return spring was small in
diameter and entered the bolt to drive the firing
pin. The rear of the spring was housed in a small
tube recessed in the cap that closed the end of
the receiver. Yet another refinement was the fitting of a sliding cover to seal the magazine opening against dirt when not in use.

The '1938A' remained in production throughout the war. As the war progressed, minor alterations, such as transforming the barrel jacket into
a pressed and welded unit, were made to allow
cheaper and easier production and to incorporate
some elements of mass production. At this stage

a new compensator design was introduced consisting of four vertical vents cut across the muzzle.

Like all other participants in the war, the Italians soon realised they were going to have to introduce much more mass production in order to keep up with demand for equipment. The Beretta answer, as far as the submachine gun was concerned, was for Marengoni to redesign the 'Model 38A' into a more 'productionised' weapon.

● **Beretta Model 1938/42** This model incorporated many of the features of the previous model but was extensively simplified to permit mass production. External appearances were changed, the stock was cut off at the magazine housing and the adjustable rear sight was replaced with simple fixed 'U' combat sight. The perforated jacket was also dispensed with in favour of parallel flutes cut the full length of the barrel to assist cooling. This was later found to be unnecessary and abandoned in favour of a smooth barrel. The compensator was reduced to two slots. Internally, the separate firing pin mechanism was replaced with pin fixed directly into the face of the bolt. In comparison to its predecessors, the 'Model 1938/42' was a utilitarian gun with welding and stamping used wherever possible in place of machining. The later, smooth barrelled version of this gun is sometimes designated the '1938/43'.

● **Beretta Model 1938/44** The final variant of the gun in which the weight and size of the bolt were reduced, which in turn led to the exchange of the small diameter return spring and its guide rod for a shorter, larger diameter spring, similar to that used in the 'Sten'. This model also reverted to the full length wooden stock and four slot compensator. All models were built by the Beretta factory at Brescia.

JAPAN

The Japanese did not begin development of their own submachine guns until 1935 although they had previously bought quantities of the Bergmann designs from Switzerland, which were rechambered for the 7.63mm Japanese cartridge and which had a bar attached to accept the standard rifle bayonet.

Top:
The Japanese were not large scale users of the submachine gun and its 'Type 100' saw only limited use issue.

Above:
A side view of the Japanese 'Type 100'.

In general, the Japanese authorities seem to have little enthusiasm for this type of weapon and those that they had bought were issued only to Marine units. This attitude is surprising as the army could have expected to have to fight in dense jungle, an environment where the submachine gun excels. The army authorities finally and rather half heartedly prevailed upon the Nambu company to design a weapon. The result was the 'Model 100', first issued in limited numbers in 1941, using the weak 8mm Japanese pistol round. It was never a successful weapon, partly because of the ammunition and partly because the Japanese seem to have been undecided whether they wanted a submachine gun or an automatic rifle.

● **Model 100** This was nearly three feet long, looking very much like a rifle and expensively machined. Overall weight was low at just eight and a half pounds and the weapon was fed from a magazine on the right side. Sights were calibrated to an exceptionally optimistic 1,500m. Great importance was given to the fitting of a bayonet lug below the muzzle, which was specially strengthened. The rate of fire was very low at only 450 rounds/min. A version was produced for use by paratroops, in which the wooden stock was hinged just behind the trigger guard. Some examples of the 'Model 100' are found with a bipod.

● **Model 100 (1944)** Although basically similar to the earlier 'Model 100', this was fitted with a weaker recoil spring which allowed the cyclic rate of the weapon to increase to 800 rounds/min. Manufacture was simplified by the bayonet bar being replaced by two lugs on the barrel casing and the sight fixed at a more realistic 100m. The Japanese submachine guns, which were built at the State arsenals, are rare weapons and probably less than 10,000 were ever built.

SOVIET UNION

The Soviet Union made greater use of submachine guns than any other combatant in World War 2. Several factors forced this upon them, chiefly being that the Germans had destroyed or captured a large proportion of their manufacturing capability, denying them vital machine tools. The Russians were therefore forced to adopt simple designs that could be produced from pressings and stampings. The Russians also had a very large army to equip very rapidly, with only time for minimal training; the submachine gun was therefore ideal equipment.

Russia in fact had a submachine gun in its armouries from the 1930s — the 'Pistolet Pulyemet Degtyarev' or 'Degtyarev Machine Pistol' ('PPD34'), a derivative of the 'MP18' that not only used the normal box magazine but also an unusual 71-round drum magazine. A good quality weapon that was used extensively in the Spanish Civil War, it fired the 7.62mm Russian pistol round but did not lend itself to mass production and, for this reason, in 1940 it was decided to redesign the weapon.

● **PPD1940** Like its predecessor, this was designed by Degtyarev. It retained its similarity to the 'MP18' but the cooling slots in the barrel casing were simplified to a few long cuts. In common with all Soviet submachine guns, the barrel was chrome plated inside for wear resistance. An alteration was made to the magazine housing to

Below:
Almost a trademark of the Red Army, the 'PPsh 1941' was an exceptionally good weapon. It used either a 35-round magazine ...

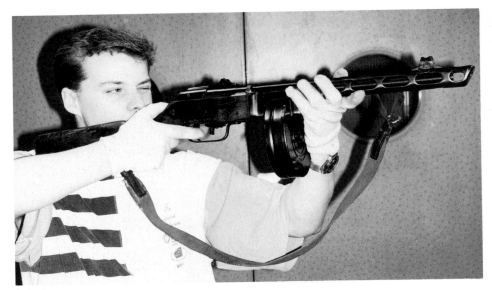

allow the drum to be inserted from the side in the manner of the Thompson. Despite simplification, the 'PPD40' was still demanding of scarce raw materials and machine tooling and, by Russian standards, few were made. The 'PPD40' was built at the State arsenals.

● **PPSh41** After much searching, a design by George Shpagin was selected as a replacement with the main stipulation that it should be easy to manufacture in quantity. In fact, construction of the 'PPSh41' could not have been much simpler. The receiver and the barrel jacket was stamped from a single sheet of steel and formed into a rectangular section and the whole assembly pinned and welded together with an extension of the jacket at the muzzle to act as a compensator. The bolt could be formed in a simple turning operation and the barrel could be obtained by halving a standard rifle barrel. The whole thing was then mounted in a crude wooden stock. The first models had a tangent sight but this was quickly replaced by a simple peep sight.

The 'PPSh41' was to become almost a badge of the Russian army, for there are few photographs of its troops in action that does not depict a 'PPSh41'. The weapon was designed to take either the 71-round drum of the 'PPD' or a slightly curved 35-round box magazine. Despite its crude construction and its high rate of fire, the 'PPSh41' was a sound gun, capable of operating reliably in even the most adverse conditions, with the minimum maintenance. During the war, more than 5 million guns were built and the design was extensively copied by Soviet satellites.

Above:
... or the more usual 71-round drum magazine.

Right:
Russian troops at Dnieper with the mainstay of the Soviet infantry — the 'PPSh 41' submachine gun.
IWM/RUS4602

● **PPS42** The Russian philosophy was, once a good design was evolved, to produce it in quantity. One standard gun was their ideal but wartime circumstances were to preclude this and bring one more Soviet submachine gun into the war. In 1942, the German army laid siege to Leningrad and stocks of arms in the city were not large enough to supply all the citizens manning the defences. With no new supplies from outside possible, A. I. Sudarev, an engineer at the Leningrad ammunition works, designed a submachine gun which could be manufactured within the resources of the city and whose design was governed purely by what materials and machinery were available.

Apart from two small pieces of wood as pistol grips and a fragment of leather as a buffer for the bolt, the entire construction of the 'PPS42' is of metal. The receiver body was stamped from one sheet of steel, the barrel jacket from another and the two then bent and crudely welded and riv-

Above:
Necessity is the mother of invention and the 'PPS 42' was developed in Leningrad (now renamed St Petersburg), during the siege using the limited materials and machinery available. Most 'PPS 42s' disappeared after the war and this model is, in fact, a later Chinese facsimile.

eted together. A piece of rifle barrel was welded in place, a turned bolt inserted and a simple folding metal stock added. The result was a perfectly adequate, if ugly, killing machine and a tribute to the skill and ingenuity of a city under siege. Once the siege was lifted, a slightly modified version was produced, incorporating lessons learned in battle and estimates claim that one million 'PPS42s' were manufactured.

After the war, the majority of these were given to satellite countries for, in many ways, the defenders of Leningrad were an embarrassment and the whole siege was played down for many years. Disposing of weapons, which would always be a monument to Leningrad's resistance, could have been a convenient way of sweeping the whole affair under the carpet. It is certainly significant that while both Degtyarev and Shpagin had honours showered upon, no such awards appear to have been made to Sudarev.

UNITED STATES

Although official credit for the invention of the submachine gun is credited to the Italians and Germans, American designer Hiram Maxim also has a justifiable claim, having produced a working submachine gun, using his 'toggle-link' principle, in the 1890s. The weapon, in 0.22in calibre, was never marketed and quickly disappeared and it was not until the 1920s that the United States had its first workable gun of the type and one that was to lead to, in 'Tommy Gun', the submachine gun's most enduring nicknames.

However, although the Americans had a thriving firearms industry, some of the world's most innovative designers and enormous experience of industrial mass production, they were remarkably tardy in adopting the submachine gun as a military issue weapon. The reason they had one at all was due solely to the forethought and perseverance of John Taliaferro Thompson, a former soldier and erstwhile gun designer, who, by accident, provided a submachine gun totally different from, and owing nothing to, those being built in Europe.

● **THOMPSON** When Thompson eventually left the army in 1914, he was determined to develop an automatic rifle for the military. His experience had convinced him that such a weapon was needed and that to be successful, it would have to be as simple as possible. For this reason, he discarded the complexities of gas and recoil operation in favour of 'blow back'. He did however

realise that the
power of the service 0.30in-'06
rifle round dictated a locking system and he
wanted something that had not been used before.
The answer was to use the 'Blish Lock' originated
for naval guns which incorporated a metal wedge
against which the bolt had to work in order to
open. Under peak pressure, the wedge would
stick into deeply inclined slots and prevent bolt
opening. As pressure dropped, the wedge would
slip out of the way, leaving the bolt free.

Thompson formed the Auto Ordnance Corpo-
ration to develop the gun but with no practical
engineering experience, he employed Theodore
Eickhoff for the practical work. Eickhoff built

Above:
The American Thompson 'M1928A1 pictured here is
a late model, produced without some of the external
frills of the earlier commercial models.

Below:
The Thompson was a large, heavy and cumbersome
weapon, particularly when fitted with its drum
magazine.

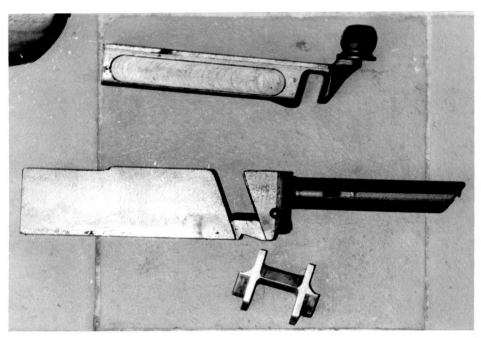

several prototypes but found that, unless each round was well lubricated, the lock would not function with the 0.30in-'06 round. The only round that would work satisfactorily was the 0.45in Colt Automatic Pistol cartridge of the US Army service pistol. Thompson was quick to see that his automatic rifle concept was doomed but equally quick to grasp the potential the pistol calibre weapon might have as a 'Trench Broom' that would sweep the enemy out of their trenches on the Western Front. Completely coincidentally, he had stumbled onto the same path followed by the Germans.

The first models were belt fed and jammed consistently. Alterations were made to the mechanism to overcome the problems and, in impatience to try the modifications, Eickhoff and his assistant, Oscar Payne, fitted a simple, spring loaded box magazine which proved so successful that all thoughts of belt feed were dropped. The prototype gun was called the 'Annihilator' and showed enormous promise. However, the Armistice in November 1918 prevented Thompson getting the gun into the action and possible rapid adoption by the army.

Rechristened the 'Thompson Submachine Gun', further improvements were made. A clockwork-powered drum magazine holding 100 rounds was developed and the mechanism was simplified. Approaches to the army produced the opinion that the gun would encourage the rank and file to waste ammunition and thus create supply problems. Despite lack of valuable military orders, Thompson persisted and promoted sales, in the process inventing the descriptive phrase, 'Submachine Gun'. The first production gun, the '1921 Thompson', was an elegant weapon with rich traditional metal finish, a detachable walnut stock and a capability of selective fire, with its rate of fire slowed from 1,000 to 800 rounds/min. It had the vertical wooden foregrip and finned barrel that were to characterise the Thompson for years to come.

At this stage the Thompson entered a phase that saved it from an early demise but which also earned it tremendous notoriety. The Volsted Act had introduced Prohibition to the United States and in turn had brought gang warfare to the streets. While the military were resistant to the

Thompson, the gangsters were quick to see its advantages and, in many States, were able to buy the weapon across the counters of sporting goods stores. While this period, which has been comprehensively, if often inaccurately portrayed in countless motion pictures, ensured the financial survival of the 'Thompson', it virtually destroyed the credibility of the gun as a military weapon and enabled its critics to dub it 'a gangster weapon'. Eventually, worn down by the notoriety his invention had gained and the apparent disinterest of the military, John Thompson retired, disillusioned, from the company at the end of the 1920s and Auto Ordnance stumbled from crisis to crisis, with only limited adoption by the US Marine Corps keeping it going to the outbreak of war.

● **Thompson M1928A1** The Thompson was finally taken into US Army service, principally for reconnaissance and armoured units in 1938 and given the designation 'M1928A1'. This model, manufactured by Colt, was basically the 1921 model, fitted with a slotted Cutts compensator on the barrel in an attempt to keep the muzzle down. A luxurious and extremely well made weapon, it retained the Blish lock, the detachable walnut stock and was fitted with a complex Lyman sight housed between two wings at the rear of the receiver. It also had the finned barrel and top mounted cocking knob. The 'M1928A1' was also ordered by the British and French governments and was often fitted with a horizontal fore grip in place of the original vertical model.

Before production ceased in late 1941, a slightly simplified version was produced in which the Cutts compensator was removed and the fully adjustable Lyman sight replaced with a non-movable 'L' peep sight. The radial finning of the barrel also disappeared.

● **Thompson M1** As with many weapons designed and developed prewar, demand for the Thompson outstripped the ability to manufacture and so a process of redesign began, leading to the adoption of the 'M1 Thompson' in 1942. The Blish lock, the built-in oiling pads, the Cutts compensator and the cooling fins were all dispensed with and the butt stock was no longer detachable. The cocking handle was moved to the right side of the bolt to ride in a groove in the receiver. As the butt was no longer detachable, for stripping, the frame was made to slide over a protruding track on both sides of the receiver. The 'M1', like the earlier model 'Thompsons' had a separate firing pin operated by a hammer. A simple fixed aperture rear sight guarded by a pair of wings was fitted on early versions although the wings were subsequently dropped. The commercial quality finish found on many early 'Thompsons' gave way to military Parkerised metal.

● **Thompson M1A1** This model differs from the 'M1' only in having its firing pin machined in the face of the bolt, eliminating the separate hammer

Top left:
Internally the early Thompsons were complex and incorporated the redundant Blish locking mechanism.

Bottom left:
The '1928A1' Thompson could take 50 and 100-round drum magazines. The cartridges were loaded between clockwork-powered rotor arms.

and firing pin of its predecessors. Neither the 'M1' or 'M1A1' Thompsons would accept the drum magazine — which had proved troublesome in dirty conditions — and a 30-round box magazine was introduced to boost capacity.

Examples of the 'Thompson' will be found carrying the maker's marks of Colt Patent Firearms Co, Savage Arms Co and of Auto Ordnance. A very few 'Model 1928A1s' were manufactured in England by BSA.

● **0.45in M3 'Grease Gun'** Despite its modifications, the 'Thompson' was still expensive and time-consuming to manufacture and, based on experience gained in the use of submachine guns, the Americans decided to start from scratch with a new design. They began with the concept that reliability and simplicity of manufacture were to be the key factors and many lessons learned in the development of the 'Sten' were

Below:
For military operations the drums were noisy and the box magazine was favoured.

Bottom:
The simplified Thompson 'M1A1', really a totally different weapon, had a straight blow back operation.

applied to the design. The result which issued from the Aberdeen Proving Grounds of the US Ordnance Corps was the 'M3' an ugly and very simple gun with an extending wire stock, short barrel and tubular receiver. Its shape immediately caused it to be christened the 'Grease Gun' for its close similarity to that familiar piece of garage equipment. There was no wood anywhere in the design. A box magazine hung from a housing, which acted as a fore grip, below the receiver. Cocking was achieved by a crank handle below the receiver. A hinged cover protected the ejection port and, when closed acted as a safety catch by blocking forward movement of the bolt. The rate of fire was exceptionally slow at 450 rounds/min and, ingeniously, a conversion kit of barrel, bolt and magazine adapter were supplied, which allowed the gun to be simply converted to fire 9mm Parabellum ammunition. Over 600,000 were made, at a unit cost of about $5, by the Guide Lamp Division of General Motors at Anderson, Indiana.

● **0.45in M3A1** In 1944, as a result of complaints about wear in the cocking mechanism, the whole relatively complex mechanism was eliminated in favour of a hole cut into the forward end of the bolt. Cocking was achieved simply by hooking the finger into this hole and drawing back the bolt. Other smaller refinements were made such as enlarging the ejection port and adding a magazine filler to the stock to produce the 'M3A1' of which some 15,000 were built. Limited quantities of a silenced version were also manufactured for the clandestine operations of the Office of Strategic Services.

Submachine gun technical data

Britain & Commonwealth

Type	Operation	Length	Weight	Barrel length	Calibre	Feed	Cyclic rate
LANCHESTER MK1	Selective	33.5in	9.65lb	7.9in	9mm	50-round box	600 rounds/min
LANCHESTER MK1*	Full auto only	33.5in	9.65lb	7.9in	9mm	50-round box	600 rounds/min
STEN MkI		35.25in	7.8lb	7.75in	9mm	32-round box	550 rounds/min
STEN MkII		33.5in	6.62lb	7.75in	9mm	32-round box	550 rounds/min
STEN MkIII		30in	7lb	7.7in	9mm	32-round box	550 rounds/min
STEN MkV		30in	8.5lb	7.8in	9mm	32-round box	575 rounds/min
STEN MkIIS		37in	7.48lb	3.61in	9mm	32-round box	Intended mainly for semi-automatic fire.
OWEN MkI	Full auto	31.8in	9.37lb	9.8in	9mm	30-round box	800 rounds/min
AUSTEN MkI	Selective	33.25in	9.2lb	7.8in	9mm	32-round box	550 rounds/min

Germany

Type	Operation	Length	Weight	Barrel length	Calibre	Feed	Cyclic rate
MP38	Full auto	32.8in	9.5lb	9.9in	9mm	32-round box	500 rounds/min
MP40	Full auto	32.8in	8.87lb	9.9in	9mm	32-round box	MM500 rounds/min
MP40/2	Full auto	32.8in	10lb	9.9in	9mm	two x 32-round boxes	500 rounds/min

Italy

Type	OPeration	Length	Weight	Barrel length	Calibre	Feed	Cyclic rate
BERETTA M1938A	Selective	37.25in	9.25lb	12.4in	9mm	10, 20, 30 or 40-round box	600 rounds/min
BERETTA M1938/42	Selective	31.5in	7.2lb	8.4in	9mm	10, 20, 30 or 40-round box	550 rounds/min
BERETTA M1938/44	Selective	31.5in	7.2lb	8.4in	9mm	10, 20, 30 or 40-round box	550 rounds/min

Japan

Type	Operation	Length	Weight	Barrel length	Calibre	Feed	Cyclic rate
TYPE 100/40	Selective	35in	8.5lb	9in	8mm	30-round box	450 rounds/min
TYPE 100/44	Selective	36in	8.5lb	9.2in	8mm	30-round box	800 rounds/min

Soviet Union

Type	Operation	Length	Weight	Barrel length	Calibre	Feed	Cyclic rate
PPD 1940	Selective	30.63in	11.9lb	10.63in	7.62mm	71-round drum	900-1,000 rounds/min
PPSh 1941	Selective	33.15in	11.99lb	10.63in	7.62mm	71-rd drum or 35-round box	700-900 rounds/min
PPS 1942	Full auto only	33.72in	7.98lb	9.45in	7.62mm	35-round box	650 rounds/min

United States

Type	Operation	Length	Weight	Barrel length	Calibre	Feed	Cyclic rate
THOMPSON M1928A1	Selective	33.75in	11.9lb	10.5in	0.45in	50-round drum	600-725 rounds/min
THOMPSON M1 & M1A1	Selective	32in	11.99lb	10.5in	0.45in	20 or 30-round box	700 rounds/min
M3	Full Auto	29.8in	8.15lb	8in	0.45in	30-round box	350-450 rounds/min
M3A1	Full Auto	29.8in	8lb	8in	0.45in	30-round box	350-450 rounds/min

Dealers

By and large, the dealers in deactivated weapons are an honest and decent bunch of people with a personal interest in what they sell and keen to provide the customer with what he wants. Therefore, if you have a special requirement, either in type of weapon or condition, make this plain to the dealer who will usually be happy to select items in particularly fine condition, if this is what is required, or to attempt to locate an item not normally in his stock. Fakes and forgeries do not as yet exist in this field of collecting and so the buyer can usually be confident that an advertised item is what it purports to be however, it is worth examining items wherever this is possible to ensure that deactivation has been conducted in a 'sympathetic' way that does not destroy the appearance of the weapon.

● **Worldwide Arms**, PO Box 5, Eccleshall, Staffordshire ST21 6SN. Telephone: 0785 851515. Mail order through regular catalogues. Also attendance at major arms fairs.

● **Manton International Arms**, 140 Bromsgrove Street, Birmingham B5 6RG. Telephone: 021 666 6066. Mail order through catalogues and sales through Birmingham showroom and arms fairs.

● **Ryton Arms**, PO Box 7, Retford, Nottinghamshire DN22 7XH. Telephone: 0777 860222. Mail order through catalogues.

● **Jeremy Tenniswood**, 28 Gordon Road, Aldershot, Hampshire GU11 1NB. Telephone: 0252 319791. Specialist in militaria including some deactivated firearms.

● **Worthing Guns**, 80 Broadwater Street West, Worthing, Sussex BN14 9DE. Firearms dealer with extensive stocks of deactivated weapons.

● **Apollo Firearms**, PO Box 1761, Potters Bar, Herts EN6 1NY. Telephone: 071 739 1616. Mail order of selected military weapons.

● **Target Arms**, 165 Lordship Lane, London SE22 8HX. Telephone: 081 693 4211. Specialist dealer in armour and deactivated weapons.

● **Uttings Gun Co**, 54 Bethel Street, Norwich, Norfolk NR2 1NR. Telephone: 0603 621776. Gun dealer holding stocks of deactivated weapons.

● **RIFLE**, 163 Alfreton Road, Little Eaton, Derbyshire DE2 5AA. Telephone: 0332 831024. Deactivated weapons by mail order.

● **Intergun**, Carvean, Probus, Truro, Cornwall TR2 4HY. Telephone: 0872 52243. A limited selection of deactivated weapons by mail order.

● **Interarms (UK) Ltd**, Interarms House, Worsley Street, Manchester M15 4LE.Telephone: 061-833 0701. The world's largest private dealer in small arms, offering a limited selection of deactivated weapons but definitely worth approaching for the more unusual weapons.

● **Selguns**, 14 Loampit Hill, Lewisham, London SE13 7SW. Offers a few selected deactivations.

● **Sherwood Armoury**. Telephone: 0227 262217 A firearms dealer offering a small selection of deactivations.

● **Horsham Gun Company**, PO Box 150, Horsham, West Sussex RH12 3FB. Telephone: 0403 60797. A limited selection of deactivations but some quite unusual items.

● **Ranger Arms Co Ltd**. Telephone: 0279 87006. A general dealer in firearms who offers deactivation of items in stock.

Glossary of Technical Terms

A conscious decision was taken to keep possibly repetitive technical descriptions from the main body of the book. However, for those wishing to delve further into the methods of operation of military weapons, the following glossary may help in explaining commonly used terms:

● **Accelerator** Used in many recoil operated weapons to provide extra velocity to the breechblock during recoil.

● **Advanced Primer Ignition** Mainly (but not exclusively) applied to submachine guns, the cartridge primer is struck and fired while the bolt is still travelling forward and the cartridge still being moved into the chamber. The rearward pressure of the exploding cartridge has first to overcome the inertia of the moving bolt before it can begin the 'blowback' (qv) action. The delay this causes allows the bullet to exit the muzzle and chamber pressure to drop to a safe level before the bolt begins to reopen.

● **Air Cooling** An automatic weapon builds up a significant amount of heat in operation and it is necessary to cool the barrel to reduce wear and keep temperatures below the level where cartridges may be set off spontaneously. A free flow of air around and through the barrel helps but, in most infantry weapons, a means of rapid change of barrels has also to be provided.

● **Assault Rifle** A light rifle designed to be capable of automatic fire and generally firing a special short cartridge. In some forces, such rifles have now taken over the role of the submachine gun.

● **Automatic** The system of operation in which the firearm continues to fire as long as the trigger is pressed and there are rounds in the magazine.

● **Automatic Pistol** A term which is generally misused to describe the semi-automatic pistol, where the power of the firing cartridge is harnessed to recock the weapon and place another round in the chamber ready to fire at the next squeeze of the trigger. There are a few true automatic pistols which operate in the manner of submachine guns.

● **Belt Feed** Method of feed for automatic weapons where cartridges are held side by side in a long belt of webbing or metal links. In some metal belts the links are held together by the cartridges themselves and, once the cartridge is removed for firing, the link is discarded individually from the gun.

● **Blowback** A system used in self-loading weapons in which the barrel and bolt are not locked mechanically. The mass of the bolt and the strength of the return spring provide inertia to keep the mechanism closed until chamber pressure falls.

● **Blow-Forward** Similar method of operation to 'Blowback' (qv) but in which the breechblock or bolt is fixed to the frame of the weapon and in which the barrel moves forward. The action is however awkward and has therefore been only seldom used in weapons such as the German 'Schwarzlose'.

● **Bolt Action** A method of closing the breech of a weapon manually in a fashion generally similar to operating a door bolt. There are also straight pull bolt actions where the manipulation of the bolt is in a straight line and the turning of the bolt achieved by cams.

● **Box Magazine** A method of feed in which cartridges are held in a metal box which is either detachable from the weapon or which forms part of the weapon's body. The magazine may contain either a single or double column of cartridges sitting on a platform which is moved upwards by spring pressure.

● **Breech** The rear section of the barrel containing the cartridge chamber.

● **Buffer** A resilient section at the rear of a machine gun, serving to arrest and cushion the movement of the recoiling parts of the mechanism. In some weapons the buffer is used to accelerate or decelerate operation of the mechanism.

● **Calibre** The diameter of the weapons barrel measured from 'Land' (qv) to 'Land' and expressed either in 1,000ths of an inch or in millimetres.

● **Cannelure** A groove in the body of a bullet into which either lubricant or the mouth of the cartridge case may be pressed.

● **Cap (or Primer)** A small copper cap filled with detonating compound used to initiate firing of the cartridge propellant.

● **Carbine** A short rifle primarily used by those whose principle function is not use of a shoulder arm (eg, artillery or engineers).

● **Cartridge Headspace** The distance between the face of the bolt and the base of the cartridge case. If not totally accurate, the cartridge may burst or the bolt be prevented from locking.

● **Chamber** The part of the rear of the barrel which houses the cartridge.

● **Charger** A metal clip used to hold cartridges and to allow rapid reloading of magazines.

● **Compensator** A device either fixed to the end of, or an integral part of, the barrel of an automatic weapon which deflects some of the exit gases upwards to counter the tendency of the muzzle to climb during automatic fire.

● **Cook-Off** Premature ignition of a cartridge by heat from the barrel of a weapon. Most light automatic weapons are designed so that the bolt is held back when the trigger is released, to allow air to flow through the barrel.

● **Cylinder** The rotating part of a revolver which holds the cartridges in readiness for firing. The cylinder is moved round, as the weapon is cocked, to align the cartridges with the barrel.

● **Delayed Blowback** Basically a 'blowback' (qv) weapon in which some mechanical means of slowing the opening of the bolt is provided.

● **Disconnector** Part of the firing mechanism of self-loading weapons to disconnect the trigger from the mechanism when a shot is fired and which does not reconnect it until the firer releases the trigger.

● **Double Action** A pistol trigger mechanism in which the user may either cock the weapon by pulling back the hammer manually or by a continuous pressure on the trigger.

● **Double Trigger** Mechanism used in some machine and submachine guns where one trigger provides single shots and a separate trigger is used for automatic fire.

● **Drum Magazine** A cylinder or shallow drum used as a magazine for a weapon. It offers the advantage of housing a larger number of cartridges than a 'box magazine' (qv) but in use has been found heavy and noisy as well as being complicated and expensive to manufacture.

● **Ejector** A mechanism to ensure the positive expulsion of the spent cartridge case from the weapon.

● **Extractor** A mechanism attached to the block or bolt of a weapon to pull the empty cartridge case from the chamber for ejection.

● **Firing Pin** A pin house or fixed in the face of the bolt of a weapon to strike the primer.

● **Fluted Chamber** Longitudinal grooves are cut into the chamber of some 'blowback' (qv) weapons, particularly those using necked cartridges. The flutes allow gas to seep around the outside of the cartridge, equalising pressure and making it easier to extract.

● **Fore-End** The part of the stock of the weapon ahead of the trigger guard.

● **Full-Cock** Situation when the mechanism of a weapon is ready to fire.

● **Gas Operation** In some automatic or semi-automatic weapons a portion of the propelling gas is tapped off through a port in the barrel to act on a piston within a cylinder. Gas pressure causes the piston to move back, operating the mechanism of the gun.

● **Gas Seal** A type of revolver in which the cylinder, or part of it, moves forward into the mouth of the barrel chamber to form a seal. .

● **Groove** Spiral groove cut into the barrel of a weapon to make the projectile spin in flight.

● **Grip Safety** A lever set in the grip of a pistol or submachine gun to lock the mechanism of the weapon in a safe condition unless the weapon is properly held.

● **Half-Cock** A situation when the firing mechanism of a weapon is partly to the rear, immobilizing the trigger.

● **Heavy Machine Gun** Machine gun of a calibre larger than a standard rifle.

● **Hinged Frame** A type of revolver in which barrel and cylinder form a unit hinged to the frame by a pivot and which swing down for loading and ejection.

● **Light Machine Gun** A machine gun fired from a bipod or similar support and intended to be controlled from the shoulder.

● **Long Recoil** System of operation of automatic and semi-automatic weapons in which barrel and breech are locked at firing and recoil, locked, for a distance that is greater than the length of the unfired cartridge before unlocking

● **Machine Gun** An automatic weapon, usually of rifle calibre.

● **Machine Carbine** Alternative name for the 'submachine gun' (qv).

● **Machine Pistol** As above.

● **Magazine** A container to hold cartridges. Sometimes removable and sometimes an integral part of the weapon.

● **Magazine Safety** A safety device for semi-automatic pistols which disables the mechanism and which will not allow the pistol to fire when the magazine is removed.

● **Medium Machine Gun** A machine gun designed to fire from a rigid rest such as a tripod.

● **Muzzle Brake** An attachment to the muzzle of a gun to divert some propellant gas sideways and to the rear to counter the recoil force.

● **Obturation** Means by which gases are prevented from escaping from the breech of a weapon during firing.

● **Primer** Another name for the percussion cap in the base of a cartridge.

● **Receiver** The main body of a firearm, housing the mechanism and to which the stock, sights and pistol grip are fixed.

● **Recoil Intensifier** A device attached to the muzzle of a weapon to trap part of the muzzle blast and turn it back against the face of the barrel to give additional rearward velocity.

● **Recoil Operation** System of operation of automatic and semi-automatic weapons relying on recoil of the barrel during firing for the reloading operation.

● **Recoil Spring** The spring which returns the mechanism of an automatic or self-loading weapon after recoil.

● **Rifling** Spiral grooves cut into the bore of a weapon to impart spin to the projectile to stabilise it in flight.

● **Safety Catch** Device to prevent accidental discharge of the weapon when cocked and loaded.

● **Sear** The part of the firing mechanism that, linked to the trigger, engages with the bolt or firing pin against spring pressure and which, when pulled clear by action of the trigger, allows firing.

● **Selector** Device which allows selective operation of the 'disconnector'(qv).

● **Self-Loading** Mechanism in which the power

produced from the discharge of the cartridge is used to reload and recock the weapon.

● **Semi-Automatic** Alternative term for 'self-loading' (qv).

● **Single-Action** Pistol lock in which the hammer must be pulled back and cocked manually.

● **Solid Frame Revolver** A design in which the barrel is fixed in the frame and in which the cylinder is usually swung out for loading.

● **Spent Case Projection** Alternative description of 'blowback' (qv) where pressure on the cartridge case pushes back the block.

● **Standing Breech** The part of a revolver frame immediately behind the cylinder and which resists the recoil of the cartridge.

● **Stirrup Latch** A latch which holds the two parts of the hinged frame revolver together.

● **Stock** The non-metallic part of a small arm (usually wood).

● **Straight-Pull** Bolt action in which instead of the bolt handle being lifted to rotate the bolt out of engagement, rotation is achieved using cams.

● **Striker** Alternative term of reference for the f'iring pin' (qv).

● **Strip Feed** Feed mechanism for machine guns which, instead of magazine or belts, uses ammunition in metal trays or strips.

● **Submachine Gun** A light one-man automatic weapon using low powered pistol ammunition.

● **Swing-Out Cylinder** Revolver cylinder mounted on a separate crane in solid frame revolver so that it may be swung to one side for loading and ejection.

● **Toggle Lock** Locking system for recoil operated weapons in which hinged levers arranged in a way similar to a knee joint lock the mechanism for part of its travel. A ramp in the receiver lifts the central joint during recoil, allowing the breech to open.

● **Water Jacket** An external casing around the barrel of some automatic weapons used to contain water and to provide cooling.

Availability & Values

Although there are exceptions, there is little problem of availability for the majority of World War 2 small arms. Indeed, some are still in service with military units and even those that are not can generally be found in surplus stocks and are therefore available to dealers and collectors. Condition of military weapons of this period tends to be generally good and most weapons can be obtained in at least 'good used condition'.

The following lists provide guide prices only for weapons in this condition, deactivated with a proof house certificate. The terms n/a is used to denote weapons that are not readily available on the market at present. This does not necessarily mean that they are rare specimens in the generally accepted sense, more often that they have not been brought into the country in large numbers. Indeed many weapons so marked may be found on the live weapons market and thus available for deactivation. It is also worth remembering in the present political climate that many interesting German weapons ended World War 2 behind the 'Iron Curtain' and have been considered rare in the West.

With the disappearance of that barrier, they may now appear on markets in the West. Weapons such as the 'FG42', which, because most were issued and lost on the Eastern front, have been considered rare and been offered at prices of up to £25,000. It would not be too surprising if many weapons such as the 'FG42', captured by the Soviets and probably stored away in arsenals since the war, surfaced now in the West.

RIFLES AND CARBINES

British & Commonwealth
Rifle No1, MkIII, SMLE £99
Rifle No3, Mk1, (P-14) £50
Rifle No4, Mk 1 £75
No4, Mk1*, No4, Mk2 Rifle No5, Mk1 £189

Germany
Kar98k £165
Gew33/40 £180
Gew33/40 (foldingstock) n/a
Gew41(W) n/a
Gew41(M) £1,100
Gew43 £585
FG42 n/a
MP43 £395
MP44 £275
StG45(M) n/a

*Italy
Mannlicher Carcano M1891 £49
Mannlicher Carcano M91/24 Carbine £50
Mannlicher Carcano M38 £50
Breda 1935GP n/a

*Japan
Rifle Type 99 (long) £175
Rifle Type 99 (short) £170
Rifle Type 2 n/a
Rifle Type 38 n/a
Rifle Type 38 (Carbine) n/a
Carbine Type 44 n/a
Rifle Type 97 n/a

Soviet Union
Moisin-Nagant M1891 £49
Moisin-Nagant M1891/30 £49
Moisin-Nagant M1910 Carbine £50
Moisin-Nagant M1938 Carbine £49
Moisin-Nagant M1944 Carbine £59
Tokarev SVT-38 n/a
Tokarev SVT-40 £190
Tokarev AVT-40 £165
Simonov SKS Carbine £165

United States
Rifle 0.30in Calibre M1 (Garand) £185
Carbine 0.30in M1 £185
Carbine 0.30in M1A1 £185
Carbine 0.30in M2 n/a
Carbine 0.30in M3 n/a
Browning Automatic Rifle M1918A2 £250

PISTOLS

British and Commonwealth
Webley & Scott Mark VI 0.455in £105
Webley Mark IV 0.38in £85
Enfield No2 Mk1 0.38in £75
Enfield No2 Mk1* £69
Enfield No2 Mk1** £69
Smith & Wesson No2 0.38in £90
Browning HP £299

Germany
P08 £275
Mauser M1932 £295
Mauser M1912 £290
Walther P38 £285

*Italy
M1910 (Glisenti) £125
Beretta M1934 £140

*Japan
Meiji 26 Taisho 04 (Nambu) £175
Taisho 14 (Nambu) £175
Type 94n/a

Soviet Union
Tokarev TT30 £95
Tokarev TT33 £95

United States
Colt M1911A1 £275
Colt Model 1917 £89
Smith & Wesson Model 1917 £89

MACHINE GUNS

British and Commonwealth
Bren Mk1 £250
Bren Mk1(M) £250
Bren MK2 £250
Bren Mk3 £250
Vickers Mk1 £395
Vickers-Berthier Mk3 n/a

Germany
MG34 £375
MG34S £295
MG34/41 £295
MG42 £550

*Italy
Breda Modello 30 £295
Breda Modello 37 £375
Fiat-Revelli 1914 n/a
Fiat-Revelli 1935 n/a

*Japan
Taisho 1 n/a
Taisho 3 n/a
Taisho 11 n/a
Taisho 91 n/a
Taisho 92 £395
Taisho 93 n/a
Taisho 96 n/a
Taisho 97 n/a
Taisho 99 n/a

Soviet Union
Degtyarev DP £320
Degtyarev DPM £250
Degtyarev DShK £450
Goryonov SGM £395

United States
Browning M1917A1 £300
Browning M1919A4 £350
Browning M1919A6 £350

SUBMACHINE GUNS

British and Commonwealth
Lanchester £110
Sten Mk1 n/a
Sten Mk2 £95
Sten Mk2S n/a
Sten Mk3 £99
Sten Mk5 £195
Owen n/a
Austen n/a

Germany
MP38 £550
MP38/40 £365
MP40 £365
MP40/2 n/a
MP41 £395

Italy
Beretta M1918 n/a
Beretta M1938A £125
Beretta M1938/42 £125

*Japan
Taisho 100/40 n/a
Taisho 100/44 n/a

Soviet Union
Degtyarev PPD 1940G £200
Shpagin PPSh-1941G £220
Sudarev PPS-42 £90

United States
Thompson M1928A1 £285
Thompson M1 £270
Thompson M1A1 £270
M3 £195
M3A1 £195

* In this price guide, few prices are given for Japanese weapons. In fact there are hardly any Japanese weapons currently available on the market, the majority have stayed in the Far East.

Similarly, some categories of Italian weapons are not readily available.

Bibliography

The most popular and useful specialist reference works are listed here, with comments, under the relevant section headings. All are thoroughly recommended.

MILITARY SMALL ARMS IN GENERAL

Ezell, Edward Clinton, *Small Arms of The World* (Stackpole Company, USA, 1977) *The* manual of military small arms. Originally compiled by the late W. H. B. Smith and irregularly revised, it is now in its 11th (?) edition. The book is a mine of technical information and provides details of disassembly/assembly of the majority of 20th century small arms.

Hogg, Ian V. & Weeks, John S. Military *Small Arms of the Twentieth Century* (Arms & Armour Press, UK, 1985) charts the development of military small arms from the turn of the century, covering each weapon and variant with a brief description. Well illustrated, the book is regularly updated to cover the latest trends in weapons development.

Markham, George. *Guns of The Reich* (Arms & Armour Press, UK, 1989) covers German military weapons from 1939 to 1945 in a very broad way but is worthwhile more for the quality of illustration than for the information it contains.

Skennerton, Ian *British Small Arms of World War Two* (Greenhill Books, UK, 1988). A book for the real student of firearms of this period, painstakingly researched and containing enormous detail on the weapons used, quantities, manufacturers and details of variants.

RIFLES

Brophy, William, S. *The Springfield 1903 Rifles* (Stackpole Books, US 1985). The 1903 Springfield was the American equivalent of the British Lee and continued in service into World War 2. Every facet of the weapon's design and production is painstakingly detailed - a valuable source of information.

De Haas, Frank. *Bolt Action Rifles* (Arms & Armour Press, UK, 1981). This book covers every major design of bolt action rifle with a history of the major types of action detailing how they work and how they were made, together with markings and specifications.

Skennerton, Ian. *The British Service Lee* (Arms & Armour Press, UK, 1982) Incredibly detailed history of the development, production and usage of the prime British service rifle from

1880 to 1980. It contains a vast amount of information useful to the collector.

Walter, John. *The German Rifle* (Arms & Armour Press, UK, 1985). A comprehensive and well illustrated history of the development of the standard bolt action rifle designs developed by the Germans between 1871 and 1945.

PISTOLS

Bruce, Gordon & Reinhart, Christian. *Webley Revolvers* (Verlag Stocker-Schmidt AG, Switzerland, 1988). A detailed account of the many revolvers produced by this famous British firm and ideal for the collector wishing to specialise. In addition it is a 'good read' and very well illustrated.

Ezell, Edward C. *Handguns of The World* (Arms & Armour Press, UK, 1982). A large and comprehensive reference book to military handguns of the period between 1870 and 1945. It deals indepth with technical development but in an interesting way and includes much detail not available elsewhere.

Gortz, Joachim and Walter, John. *The Navy Luger* (Lyon Publishing International, UK, 1988) A classic and well researched reference book to what is probably the most important variant of the famous Luger pistol.

Hogg, Ian V. *Military Pistols & Revolvers* (Arms & Armour Press, UK, 1987) A well illustrated book which deals concisely with the history of the world's major military handguns through the 19th and 20th centuries. Not particularly a technical book, it provides a narrative account of developments.

Hogg, Ian V. and Weeks, John. *Pistols of The World* (Arms & Armour Press, UK, 1983) This is an alphabetical encyclopaedia of modern pistols, military and civilian; by no means comprehensive but a useful source of basic data.

Konig, Klaus-Peter and Hugo, Martin. *Service Handguns* (Batsford, UK, 1988). Lists and illustrates more than 200 pistols that have seen service with the police and military. Photographs of each side of each weapon provide details of markings and design while 16 items of key data, such as dimensions and date of introduction, are listed.

Stevens, Blake.*The Browning High Power Automatic Pistol* (Collector Grade Publications, Canada, 1985). A very detailed and extremely well researched book which follows development of this important handgun from its origins in 1921 through wartime production, by both the allies and in occupied Europe, to copies used by the Argentine forces in the Falkland Islands.

Walter, John. *The Pistol Book* (Arms & Armour Press, UK, 1989). A concise one volume guide to handguns, both military and civilian, organised in alphabetical order. It is very good as a catalogue of products and producers, although short on information for the collector of deactivated guns.

Walter, John. *The Luger Book* (Arms & Armour Press, UK, 1986). An encyclopaedic volume for the collector of this marque, providing every scrap of information that might be needed, from the different variants to the many different marks found on the guns themselves, as well as on accessories and even the ammunition.

Wood, J. B. *Beretta Automatic Pistols* (Stackpole Press, USA, 1985). A very clear yet detailed chronological history of the Italian service pistols. which otherwise have received little attention from writers on small arms.

SUBMACHINE GUNS

Hobart, F. W. A. *Pictorial History of the Submachine Gun* (Ian Allan Ltd, UK, 1973). The word 'Pictorial' in the title gives the wrong impression for, although well illustrated, it is not just a picture book and contains much useful information on the characteristics and development of the submachine gun. It covers weapons of all the major powers, their history and employment.

Nelson, Thomas B. & Lockhoven, H. B. *The World's Sub Machine Guns 1915-1963* (Arms & Armour Press, UK, 1983). Provides a highly detailed country-by-country survey of all weapons produced in this period with technical data and basic history.

MACHINE GUNS

Dugelby, Thomas B. *The Bren Gun Saga* (Collector Grade Publications, Canada, 1987). Authoritative and highly detailed book on this important weapon that contains all the information that even the most demanding collector could require. It even contains reprints of official users handbooks for the major marks.

Goldsmith, Dolf L. *The Devil's Paintbrush* (Greenhill Books, UK, 1989). Basically a biography of Sir Hiram Maxim, who was effectively the inventor of the modern machine gun, but also an interesting book as it interweaves the history and development of machine guns used by the major powers which incorporate some of the principles invented by Sir Hiram.

Hogg, Ian V. *The Complete Machine Gun, 1885 to The Present* (Phoebus Publishing, UK, 1979). This book details the development of automatic weapons — both submachine and machine guns — and contains lots of interesting historical fact interspersed with excellent information.

Index

Afterword

The world moves on and since the draft of this volume was prepared, the market for deactivated weapons has changed considerably.

I mentioned in the earlier text that the lowering of frontiers in Eastern Europe could release previously unknown stocks of some of the hitherto rare weapons, particularly German ordnance captured on the Eastern Front. This has since been bourne out and we are now beginning to see quantities of the once scarce 'Gew 42' and 'Gew 43' rifles being offered on the deactivated market. It is impossible to gauge in just what sort of quantities these may be available. It is also at this stage impossible to say if quantities of the extremely rare 'FG42' will appear; the majority of these were issued in the East and lost there, so there could be many thousands still hidden away in warehouses in the former Soviet Union.

Some quantity figures being mentioned for weapons from the former Soviet Bloc are enormous, for example, although not within the scope of this book, there are something like four and a half million 'AK47s' being offered on the market at the moment. We certainly will not see such quantities of Nazi weapons being offered but clearly, if weapons appear in quantities, prices will fall.